Approaching Singularity

THE GENESIS OF CREATION

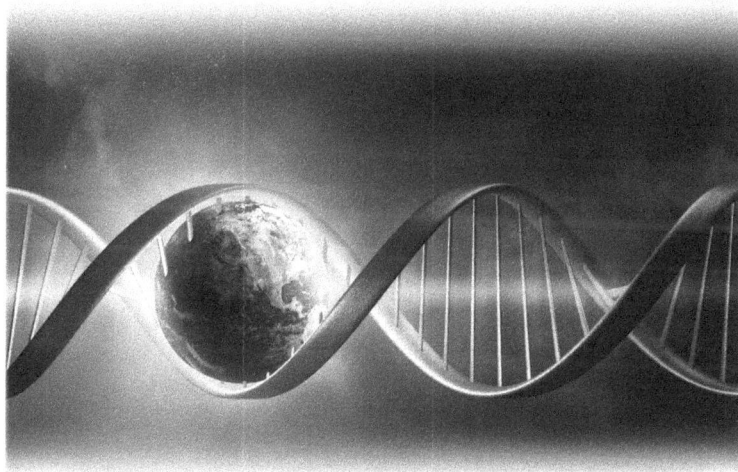

S. Jason Cunningham

Contributions and Introduction by
S. Lane Cross

Book design: bookdesign.ca
Published by:
Reinhardt & Still Publishers (**R&S Publisher PIC**)

S. Jason Cunningham
Approaching Singularity:
The Genesis of Creation
www.smokeandmirrorsbook.com

Library of Congress Cataloging-in-Publication Data
Cunningham, S. Jason
Approaching Singularity: The Genesis of Creation / S. Jason Cunningham
p. cm.
Includes bibliographical references and index.
ISBN-978-0-9885483-1-2
1. Science 2. Consciousness 3. Prophesy 4. Atlantis. I. Title – Includes bibliographical references.

Printed in U.S.A.

Reinhardt & Still
Publishers

About the Author

S. Jason Cunningham has spent the last 12 years working extensively in Southwest Asia, the Middle East, and North and East Africa specializing in conflict resolution, geopolitical research and investigations, and corporate risk management and due diligence in such countries as Iraq, Saudi Arabia, U.A.E., Libya, Uganda, Pakistan and Afghanistan. The author's career began some 19 years ago working on a diplomatic dialogue in support of the Middle East Peace Process, and has since travelled the world visiting more than 60 countries. The author has hosted high-level diplomatic meetings with Presidents, Prime Ministers and leaders of multinational corporations, and frequently speaks at international conferences on homeland security, and oil and gas. Meet the author – www.smokeandmirrorsbook.com

Also by:

S. Jason Cunningham

Smoke and Mirrors

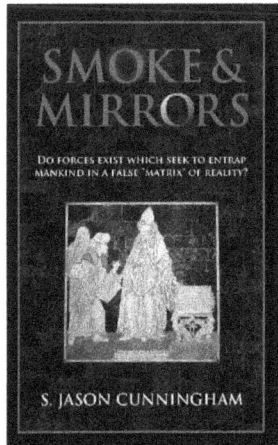

Approaching Singularity:
The Genesis Of Creation

"In this sequel to *Smoke and Mirrors*, more "dots are connected" and key questions are addressed with rational clarity and spiritual depth about the forces acting upon all of us as our world makes its next passage through the galactic plane of the Milky Way. This compelling narrative looks back into history and then points the way ahead for our mind and spirit to ascend through the coming cosmic singularity without fear or hesitation. *Approaching Singularity* will empower humanity for the next phase of our evolution."

RADM Charles R. Kubic, CEC, USN (ret)

"In her first book *Smoke And Mirrors*, the author walks us through a historical financial timeline, introduces us to our oppressors and what they have done throughout history to keep their power over us in an oppressed fear-filled state. Now, in *Approaching Singularity – The Genesis of Creation*, we learn the truth about what our oppressors desperately do not want us to discover. It is a must read for any theorist, whether you are interested in physics or religion, or anything in-between, this is a marvel in the power of thought and manifestation! A true 'one sitting' read and an awakening beyond yesterday's comprehension."

Spencer L. Cross

DEDICATION

I lovingly dedicate this book to my teacher of wisdom and truth, Jesus, the Christ

"THE LIPS OF WISDOM ARE CLOSED, EXCEPT TO THE EARS OF UNDERSTANDING."

Contents

Part 1. Approaching Singularity

Part 2. The Ascension

APPROACHING SINGULARITY

Preface

The three dimensional reality within which we live as humans evolves continuously under the influence of physical, mental and spiritual forces which shape our body, our mind and our soul. It is easy and acceptable to think of our body, our mind and our soul as evolving on separate paths subject to separate physical laws, mental boundary conditions, and spiritual beliefs. But is it possible that all three forces are actually combined into an ever-accelerating cosmic dance that only becomes obvious to us as we approach a point of cosmic singularity?

Mathematically, the simple non-linear hyperbolic equation of $XY=1$ plots as two separate opposing hyperbolic curves in the top right and lower left quadrants of a standard graph in two dimensional Cartesian coordinates. In such a graph, two points of singularity are defined, one point as X approaches positive and negative infinity while Y approaches zero, and the other point as Y approaches positive and negative infinity while X approaches zero. Does this simple mathematical model with X and Y variables also reflect the genetic basis of life and the duality of the universe? "As above, so below, as below, so above."

Physical phenomena can be defined with mathematical precision, boundary conditions which create our reality can be established mentally, and gaps between the physical and mental can be filled spiritually. But, is it possible we are all quickly approaching a point of singularity within the time-space continuum? And as we approach singularity, are physical, mental and spiritual forces actually becoming inter-twined like the DNA which defines our humanity? And are these inter-twined forces approaching infinity simultaneously as our remaining time approaches zero?

As these forces evolve and spiral around us with increasing speed while our world crosses the galactic plane of the Milky Way, will we be ready for our body, mind and soul to ascend through the approaching singularity? Or will we find ourselves trapped in a self-created reality, faced with repeating another cosmic cycle within the infinite rhythm of the universe?

Building upon the concepts which the author presented initially in her book *Smoke and Mirrors*, these and other intriguing questions are addressed with rational clarity and spiritual depth in *Approaching Singularity: The Genesis of Creation*.

RADM Charles R. Kubic, CEC, USN (ret)

———•◆•———

Introduction

The Nature of Time
By Spencer Cross

The perception of time has been described by philosophers in creative parables, and by scientists as rational mathematical formulas. However, as we approach the hyperbolic ascent in consciousness one learns that time is neither a linear formula, which can be explained through mathematics, nor an esoteric parable as defined by philosophers like Plato. Time is both a linear progression through space as identified by cycles and rhythm as well as a moldable and morphic energy, which is marked by consciousness and decisive events. Together these seemingly opposing ideals present a metaphorical song and dance through the universe waiting for an audience to watch and listen to its drama. Within this cosmic drama the message seems simple: pay attention to the present for the present holds all the secrets of time and space.

The present contains our perception of the past as well as a vision of the future. The drama is trying to tell us that we have access to all three phases of what we call time: the past, present and future collectively rolled up like a carpet ready to be transported. Where is this carpet of time to be taken? Right here, right now. The compression of time is perfectly packaged in "the now" and is ready to be experienced and accessed by everyone.

Physicists spend their energy defining space and time as a contract with certain laws, which must be adhered to and yet so many questions exist around why certain particles and certain wavelengths seem not to respect the laws and contracts. These anomalies exist and refuse to back down in defiance of the greatest scientific minds the world has ever known; almost to mock and jest the theorist. No matter how magnified the reality becomes whether it is a cell, an atom, an electron, a quark or any other infinitesimally small particle, the same result occurs – the laws are still broken and the unanswered questions continue to remain in the defiant aftermath.

Is time guiding consciousness through the universe, or is consciousness guiding time? In the compact space within these pages you will begin to learn the answer to this question. Although it may be impossible to decode time and its restrictions, a benevolent force is aiding us in our quest for answers. A reality, which seems impossible, now through internal discovery may present a flash of cosmic wisdom.

What you will learn while discovering the Approaching Singularity is that cycles foretold by the ancients define existence in both the physical and ethereal dimensions. We as a species are approaching a moment of self-awareness and self-actualization unlike any in our linear history. We are racing towards the ultimate discovery, who we really are and where we are going. Can this be the moment where our rational mind unifies with our infinite self? As you journey through this book you will join with the cycle and break free of the illusion. Enjoy your journey for it is the reason for being.

———•◆•———

PART ONE

Approaching Singularity

"Many people would claim that the boundary conditions are not part of physics but belong to metaphysics or religion. They would claim that nature had complete freedom to start the universe off any way it wanted. That may be so, but it could also have made it evolve in a completely arbitrary and random manner. Yet all the evidence is that it evolves in a regular way according to certain laws. It would therefore seem reasonable to suppose that there are also laws governing the boundary conditions."

Stephen Hawking

As we approach this epoch in humanity's evolution, or as the Mayans claim the 5th cycle of life, we find history (time) and consciousness (eternity) meet one another at the event horizon and approach a Point of Singularity.

History in the third dimensional space, in which we physically reside, is seen in the waves of time and can seemingly be mapped by mathematical formulas predicting the ebb and flow of life, and the rise and fall of civilization. Consciousness existing outside the boundary of the third dimension, hence time, brings a level of uncertainty into this mathematical formula.

Consciousness and time seemingly merge and intuitively connect to become co-creators of the material realm, hence reality, within the third dimension. This metaphysical, cosmic dance between consciousness and time brings with it the cause and effects of life as can be viewed within the world we live throughout the inevitable evolution of humanity.

It is this paradox that confounds scientists who feverishly try to tame the wilds of the universe within their mathematical formulas, precepts and experiments. Similarly it is just this paradox that brings science, religion and spirituality into unique alignment. You may argue that this could not be the case, as scientists are often atheists and share no similar concepts with either religion or spirituality. However, as you explore the Genesis of Creation and the Approaching Singularity it will be evident that in fact their varying arguments point towards the same logical conclusion. This is a truth that resonates deep within our Soul.

The Science of Miracles

"Science investigates religion interprets.
Science gives man knowledge, which is power,
religion gives man wisdom which is control."
                    Martin Luther King, Jr.

"Question everything" has become the new catch phrase in mainstream science, which seems to be a rather cool and hip approach to a cold and unfeeling industry. By its very nature, spirituality imbues the concept of miracles as an explanation for 'that' which happens and cannot be explained. Within the realms of religion and spirituality, miracles are an accepted phenomena. Could it be that the foundation of scientific theory explaining the existence of matter is also a "miracle"?

The scientific community wants us to believe that miracles simply do not and cannot possibly exist. However as you analyze the Big Bang Theory it is evident that scientists do in fact believe in miracles and apply this phenomenon to their cherished precepts.

The Big Bang Theory attempts to explain how the universe developed from a dense state of plasma, and through the sheer force of energy, albeit creative and intelligent energy, the universe began to expand over billions of years. The Big Bang Theory is essentially the moment of a point of singularity from which everything came and then expanded outwards manifesting matter and life. Not unlike the miraculous cellular explosion that happens at the moment of conception in the womb.

Science will tell you it is a unified force, a singularity, some 13+ billion years ago that 'miraculously' came into being from nothing, and in its creative energy 'miraculously' expanded into everything. This theory of nothing suddenly creating everything is perhaps the most surprisingly accepted theory within the scientific community. And yet, it is the only theory science is able to put forward to explain life and matter in the third dimension, and perhaps all perceived dimensions.

Will it be within the uncharted waters of quantum physics that the truth of humanity's eternal consciousness and being is proven? The quantum, eternal consciousness that is you and I, can be seen in the expressions of matter and life on earth. This is the tangible nature of reality in which science and spirituality are beginning to coalesce.

As the dogmas of the old fall away, the truth about the nature of our reality emerges like rays of sunshine breaking through the fog. What seems impossible to see is in fact right before our eyes. The institutional concepts of old paradigms must be shed. As we stand at the edge of knowing and lift the veil to the unseen, a quantum leap in conscious evolution is before us. Understanding where we came from and who we are is the first critical step in our collective awakening in this epic transition. And, as we collectively awaken, this transition accelerates with the momentum our energy (collective consciousness) brings it.

> "You cannot learn anything if you already feel
> that you know. Preconceived ideas and preju-
> dices always prevent us from seeing the truth.
> You should open your mind before you open
> your mouth." Master Zen

This paradigm shift is the pivotal point by which we transit towards singularity and return to the Christ Consciousness, meaning the collective consciousness of humanity in an awakened state of being in the light of God.

The logical question is, "what is singularity"? This is one of the fundamental questions we will seek to discover in this book, and it is the journey I want to take you on.

Scientists, in particular physicists, will tell you that they don't know the answer to this baffling question. But they suggest that singularity is believed to exist at the core of a "black hole" a place of immense energy and gravitational pressure that is perceived to be so intense that neither matter nor time exists, rather it is a seemingly inter-dimensional space of creative energy. Scientific theory suggests that the immense energy condenses matter into "infinite density", which is a physical return to source energy.

This certainly is a concept that boggles our minds even today, and the leading scientific thinkers of past and current generations have not yet been able to understand this extraordinary phenomenon. Thousands of years ago, a mathematically and astronomically advanced civilization in the Mexican Yucatan peninsula and highlands of Guatemala knew what physicists have only begun to understand in the last 100 years.

As far back as 1,300 BC, and some historians point to a much further date in time as I will later reveal in this book, the Mayan civilizations relentlessly studied the heavens above and their community revolved around the importance of time, its cycles and its causes and effects to life on Earth. The Mayans knew that in the center of our Milky Way Galaxy was a black hole, which they called the Dark Rift. The Mayans knew this was a force of such immense power that it was to be perceived as the source of the infinite energy of the Hunab Ku (meaning the singular creator of everything and that which gives measure and motion – God). The Mayans believed that this force was so strong that it literally pulled our galaxy around it, and the effects of this galactic dance resulted in the cosmic cycles of time and evolutionary processes on Earth.

I believe this is metaphorically understood in writings from the Gospel of Thomas, one of the 12 Apostles and half-brother to Jesus the Christ, as he writes, "In the Round Dance of the Cross, Jesus asks humanity to join in **the cosmic dance**: *"Whoever dances belongs to the whole.' 'Amen.' 'Whoever does not dance does not know what happens.' Amen"*[1]

As scientist are just now discovering, what the ancient Mayans and Egyptians already knew is that the physical crossing of our solar system and sun through the galactic plan of the Milky Way galaxy resembles the physical Cross. This is an event, which happens roughly every 13,000 years, and heralds an opportunity for the return to Source energy. What makes this cycle so unique at the end of 2012 is that we actually begin our galactic transit back through the plane of the Milky Way galaxy thus completing a full 26,000-year cycle. As we complete a full transit, our sun will align with the center of the Milky Way in a momentous event not witnessed by humans for 26,000 years.

[1] Gospel of Thomas

Thus as eloquently stated by Jesus, the "Round Dance of the Cross" illustrates the revolution of our solar system in the macrocosm of our Milky Way galaxy, which is in fact the cosmic spiral and its cyclical 'round dance'. I contend that this physical cycle mirrors the spiritual cycle in the opportunity for evolutionary advancement. As Jesus warns us, *"Whoever dances belongs to the whole.' 'Amen.' 'Whoever does not dance does not know what happens.' 'Amen.' "*[2] This is a warning to humanity to awaken to the Point of Singularity in which we approach, which Biblical text and ancient prophecy refer to as the Second Coming – the Christ Consciousness.

In returning back to modern science, similar to the Mayans, we theorize that our universe is infinitely dense and results from a singularity. The Mayans knew thousands of years ago what this singularity was and its importance for humanity. This is a message similarly conveyed by Jesus, the Christ. Amazing as it sounds, it is only in recent modern times that scientists have come to the same conclusion about the physical Round Dance of the Cross.

What ancient civilizations knew, which we are only now rediscovering, is that this physical, galactic return to source, mirrors our spiritual (soul) return to source. As a wise, ancient hermetic axiom states: "As above, so below, so below, as above". Meaning that which happens in finite form (matter), is also that which happens in eternal, infinite spirit (consciousness). The two are interlinked in a cosmic dance of evolution towards singularity being given the opportunity to return to its source of creation.

———•••———

---

[2] Ibid.

The Science of God

"All religions, arts and sciences are
branches of the same tree."

Albert Einstein

P hysicists are the meta-physical spiritualists of the scientific community. Physicists will tell you that their work shows that the only logical explanation for the inner workings of energy is to view it as acting upon a unified intelligence and as an interconnected web functioning as part of the whole. The causes and effects of energy are not simply random events, but clearly illustrate intelligent design.

An interesting choice of words is used in the world of physics. "Intelligent Design" is also the word used by religious and spiritual believers defining the supernatural and omnipotent force that is God. Could it be that as science advances and seeks to explain the unexplainable, (i.e. the unseen forces controlling matter), it is leading the sea of scientific consciousness towards the only logical conclusion – proving the existence of God?

It is within the realms of creation and the unseen power that governs energy, hence life, in which science and spirituality begin to coalesce. Where science fails to complete the equation, spirituality and the unseen mental dimension fills this gap and even gives this infinite power a name. As I explored the realms of science and spirituality, and their theories of creation, it is quite apparent that both camps of thinking have much more in common than they would like to believe. Indeed creation is by intelligent design originating from a unified force we call God.

This discussion brings us into the broader theme of this book —where did we come from and where are we as a species going? As we move through this full transit and approach the crossing of the plane of our galaxy, could the next stage of evolution be upon us?

Imagine for instance the miraculous life of a butterfly. The metamorphosis of an insect evolves from an egg to larva (caterpillar), to pupa, and finally into its spectacular transformation into adulthood as a butterfly. In the concrescence of its evolution, the butterfly transforms from simplicity into a creation of vast beauty. As you gaze on a butterfly its dazzling display of colors and graceful movement from flower to flower, it is hard to image that it once was a caterpillar scurrying up a leaf feeding and looking for shelter from other hungry insects or birds.

It's easy to see the similarities between the life cycle of the butterfly and the last 26,000 years of humanity's evolution. Like the caterpillar, our DNA is pushing us towards a metamorphosis, and like the caterpillar we must accept our fate and trust in the intelligent design of creation. The laws of nature, God, deem it so. All species must ultimately accept their evolutionary paths or cease to exist. It is God's built in safety net to ensure life rises upward and progresses, rather than digresses or de-evolves and self-destructs.

This moment in time is widely anticipated by so many, and prophesized by ancient civilizations and religious prophets. This is the moment in time within our evolutionary cycle that we will have the chance to transform from the veraciously hungry caterpillar into the spectacular butterfly and take flight.

As our beautiful solar system and Earth makes its galactic transit across the plane of the center of the Milky Way galaxy and

returns to the source creation of matter, our souls mirror this physical transit and will soon emerge from the cocoon of their slumber
to awaken to a re-birth of evolved consciousness.

The Butterfly Effect

*"What the caterpillar calls the end of the world,
the Master calls a butterfly."*

Richard Bach

One of the first lesson's in biology we teach our children is the miraculous story of the butterfly. The eggs of the adult female butterfly are laid in the Spring, Summer or Fall, and the butterfly will lay many eggs not knowing how many of her offspring will survive and fulfill its transformative destiny into becoming a butterfly. The parallels of God, human consciousness and our soul are quite striking.

As the egg hatches and becomes a larva, caterpillar, its objective is to grow and consume as much as possible as it slowly transforms its body shedding its skin 4-5 times and growing 100 times its initial size. As the caterpillar reaches its evolution and has sated its physical needs it becomes a pupa or chrysalis. The chrysalis suspends itself in a cocoon of silk where its DNA triggers an extraordinary transformation, and from the cocoon emerges the dazzling butterfly, which takes flight.

The Mayans say that the last 5,000+ year cycle of humanity has been the age of the "People of Maize", meaning the age in which mankind sought to grow through consumption and acquisition of the material. This is a period of great challenge being fully immersed in the illusion of the material dimension, a place in which many souls are dormant in their unconscious state.

Does the caterpillar know at birth that it will become the spectacular butterfly? Our logical mind says probably not, its evolutionary path designed by God, is not truly known to the caterpillar until it 'awakens' from its cocoon. Similarly, many ancient civilizations and messengers of God, which I and spiritualists call "Ascended Masters" (ascended meaning already awakened in Christ Consciousness), tells us that we must awaken to the physical and spiritual cosmic dance and thus realize our dream of metaphorically taking flight. You can see why the Mayans call this the coming of the Golden Age and the rebirth to truth and spirituality.

Following the prophetic date of December 21, 2012, the Mayans say that we move into the age of the "People of Honey", and leave the challenging age of the "People of Maize". They poetically state that within this transition will be a time when "all will be revealed and we will remember who we are". It will be the metamorphosis of the human consciousness evolving from material consumption into spiritual rebirth and a return to collective consciousness, the Christ Consciousness.

Over the last 7,000 years of recorded human history, humanity has evolved through its various stages of growth and consumption of the material world. We find ourselves on the precipice of choice. Do we continue along the path of veracious consumption or succumb to the will of God, and awaken to our true destiny? The cycles of time, pre-determined in the evolutionary stages within the material and spiritual dimensions, give us no choice but to surrender to this wonderful inevitability.

Those souls who are not ready for this moment face the challenging reality whereby they may devolve into the lower dimension of existence, and have to repeat this 26,000-year cycle. Either choice is not to be judged, as our destiny has already been chosen by us, and God.

Evolutionary Complexity

"We are shaped by our thoughts; we become what we think. When the mind is pure, joy follows like a shadow that never leaves."

Buddha

Evolutionary complexity is the process by which matter and life evolves in its intricate processes. The complex and diverse nature of this process illustrates the intelligent design of the Universe and the cycles of time that demand transformation and upward evolution.

It is upon the platform of complexity and diversity of life that the universe evolves as matter, and the consciousness that creates it expands it through evolution. It is this platform of previously achieved complexity in which there arises new forms of ever evolving diversity. Matter is based on complex chemistry, but it is also based on conscious (the unseen eternal force). It is out of both the physical and conscious complexity that there emerges an intelligent evolutionary pattern, which is governed by cycles of time. The wave of evolution and the ebb and flow of time coalesce in artful movement, similar to martial arts such as Tai Chi, mirroring the movement of our solar system and galaxy. This is the ultimate cosmic dance.

The universe both makes and preserves this evolutionary process and records it in what appears to be a magnetic matrix of resonance sealed within time (history). This archive of earth and human history, also known as the Halls of Amenti, Akashic Records, or Hall of Records, does in fact exist, but not in this lower vibrational frequency (dimension). This is the place which ascended masters frequent, and those humans who are awakened to the higher vibrational frequencies may also visit albeit in their dreams or through meditation.

The individual advancements in evolution seemingly proceed more quickly than the stage that preceded it. As we evolve faster and faster, not as spectators to the evolutionary process but rather as conscious co-creators in the evolutionary process, this evolutionary advancement rapidly increases like a divinely predetermined countdown towards the point of singularity and event horizon.

Humans are indeed conscious players in the cosmic drama and hold a leading role in directing this evolution through our collective consciousness and free will. Each stage of advancement seems to happen faster than the stage that preceded it. This acceleration of time is obvious to us in our own lives. This isn't a perceptional illusion but actually happening to the space – time matrix as we race towards a singularity. The greater question is why? Why a quickening of time? Is this a divine law of nature that this acceleration is built in?

This is not some random runaway freight train of nature, but rather a controlled evolution of both matter and consciousness. Exploring this question is to explore the very nature of consciousness and how it directs, and in fact co-creates, the evolution of matter in this third dimension.

Surprisingly the acceleration of the 'uncontrollable natural world', and its links to the conscious co-creation expressed in free will serve to ensure that this acceleration towards singularity both happens, and that we feeble minded humans don't mistakenly perceive ourselves to be "God". We are both in control of the process, but yet we are not. The laws of nature, or more importantly the divine laws of creation, supersede our own desires, and thankfully so. Humanity must first earn its evolutionary stripes, and transition from the adolescent stage of the hungry caterpillar into the maturity of the majestic butterfly.

———•·—

The Nature of Consciousness

"You are an explorer, and you represent our species, and the greatest good you can do is to bring back a new idea, because our world is endangered by the absence of good ideas. Our world is in crisis because of the absence of consciousness."

Terrance McKenna

Our creator, God, has given humans the gift of eternal, infinite consciousness – the quantum eternal existence of our soul. Except for the rare few ascended masters who incarnate in this dimension, most people lead their lives immersed in the illusion of this material dimension not realizing their full potential in an awakened state of consciousness. Even though most souls remain dormant trapped in the physical body, the inherent consciousness breaks through our human thought processes; some call our sub-consciousness, and does affect our choices in life and evolutionary path.

What is well understood by ancient civilizations, and current day mystics, but forgotten to most of modern society is the effect of human thought and consciousness in creating the causes that bring about the effects that are visible in the third dimension. Human thought, like the elusive monad (meaning the thing at the end of time and the totality of all being) does design the world in which we live. Collectively our thought can raise a civilization to great and wondrous heights, or devolve a civilization into total despair and collapse. In every aspect of our God given free will, we are the directors of this movie we call "life".

Have you ever considered that thought and consciousness vibrating in the etheric realms is the 'thing' that creates matter? Meaning the vibratory patterns of consciousness existing within the harmonic resonance of energy, when coupled with light (photons – energy) and intention (consciousness) becomes physically manifested matter in the dense third dimension.

Most humans, not understanding this process in the third dimension (third density material world) can not control this process on an individual level, and therefore find themselves the victim of the causes that bring about their affects. Most people understand this as karmic forces. But, there have existed spiritual masters who from time to time have walked the Earth and were capable of imbuing this reality simply due to their 'awakened' state of being and their understanding of the true existence of their soul and the nature of creation.

As a collective consciousness, humans do in fact manifest those realities on Earth simply by perceiving them to be that way. The power of the collective consciousness can in fact be measured as scientifically as one might measure energy. This is the fate to which humanity has entrapped itself, and this is the fate from which we, the butterfly, will emerge.

Many ascended masters (the messengers of God) have come to Earth to show humanity 'the truth, the way', but we unfortunate souls blinded by the illusions of the material world, remain in the realms of the 'unconscious'. In the most egregious manner, some souls knowingly turn away from the light in pursuit of power in the material world, and thus manipulate and corrupt these truths for their self-gain despite the suffering of the masses. Sadly, we see this reality clearly evident in our world today, and still under the control of a limited number of so-called 'elite' families. Their journey to the light will be an extraordinarily difficult path, as they knowingly turn their back on God. This reality brings a better understanding to the ancient term, "antichrist", meaning the "unconscious", the antithesis of Christ, with Christ meaning an "awakened or conscious" state of existing in the true light of God.

In my own self-discovery I've come to learn the basic tenants of creation in the material world. As the Bible states, *"In the beginning there was the Word and the Word was with God..."*, but what could this really mean?[3]

I believe the "Word of God" is the cosmic dance of God's eternal, infinite consciousness combining with harmonic resonance (sound) to manifest matter and the physical world. Thus harmonic resonance (sound) and energy in the form of light (consciousness) are the key ingredients for manifestation. This is the metaphorical elixir of God.

Humanity, in awakening their soul, can return to this perfected state of existence while in material form. This is the magic and gift shown to us by Jesus, the Christ.

Unfortunately the male dominated dogmatic religions can't show you 'the way', as these truths have been distorted through centuries of translation, re-translation, and political, economic and social manipulation and control. But yet, as God deems it so, we are re-discovering these truths with vast speed as we approach the creative energy in the Hunab Ku.

> As stated by Terrance McKenna, a modern scholar pondering the depths of human consciousness and the I-Ching, "The modern predicament where science and spirituality coalesce is in the abyss of consciousness." The 4 abysses of which McKenna pondered are;

---

[3] Genesis 1:1

the Biological Abyss represented by death and dying is the central crisis of every individual's existence. The Historical Abyss represented by the end of history – the apocalypse appointed to our world by western religions. The Psychological Abyss represented by dreams and visions – a casting off of the ego and being at the mercy of the collective consciousness or over-mind of humanity, and lastly the Physical Abyss which is the abyss of space that surrounds this planet and the source of cosmic human isolation in the third dimension. The Physical Abyss represents the material nature, and within it the abyss of the language which casts nets against the four gulfs of the aforementioned abysses.

As McKenna explains, unifying the opposing abysses requires the integration of consciousness residing in both a parallel universe, called the mental dimension, and in the parallel continuum of the physical third dimension. Humans have taken command of the physical world, i.e. the atomic world, but this best model of reality in science does not answer consciousness, the individual and collective over-mind experience.

The problem of extra-terrestrial contact is a good analogy of the essence of the problem. Science tells us that we are alone in the Universe because otherwise there would be evidence of extra-terrestrial life, or why haven't we picked up any of their radio frequency contact.

The irony of this argument is that humans for thousands of years have communicated with other beings and had both physical and psychological experiences with non-earthly beings, which reside in alternate dimensions.

The third dimension is building on previously achieved complexity, and is speeding up towards a point of singularity. The third dimension is evolving within itself to further complex systems, which evolve at a rapid rate. If this is happening then inevitably a time will come with the rate of complexity is happening so fast it will reach an 'omega point' at the end of history. We are in the shadow of this transition, an apocalyptic time.[4]

Although McKenna died roughly 12 years ago, he was ahead of his time in understanding that from this transformative experience a new model of time (history) based upon a "real intuition" that time is in fact speeding up and that humans are somehow magically apart of this process. It is the culmination of this process that McKenna rightly points out, does in fact lead to a point in linear time in which this singularity will be achieved.

Based on McKenna's research using mathematical formulas to postulate the ebb and flow of civilizations and time, the math drove him towards the illusive end point of this singularity – the point in linear time where this event may begin to initiate humanity into a yet unknown evolution. McKenna, without having studied the Mayan prophecy and while using the I-Ching as a reference within his mathematical calculations, came to the prophetic date of December 21, 2012.

---

[4] Terrance McKenna Speech at University – date unknown

McKenna in his purely scientific analysis of this phenomenon mathematically concluded that at the end of 2012 a momentous event(s) in human history would unfold. It is no surprise that this coincides with the many ancient prophecies and the physical/spiritual "Round Dance of the Cross" of which Jesus spoke.

Is this the end, as many religions contend, or rather as I propose, a beginning? Many call this moment the Biblical apocalypse – "the end times", however even this has been misinterpreted throughout history. Apocalypse literally translated in ancient Greek means an "uncovering or revelation of something hidden". The term was traditionally used to signify the meaning of that which is "hidden from human knowledge in the era of falsehood and misconception". To this affect the word apocalypse holds the same meaning for this moment as declared by the ancient Mayans "...it is the lifting of the veil, all will be reveled and we will remember who we are".

Human evolution over the last 2,000 years of recorded history can be seen as evolving through the expression of our mind and consciousness, as manifest in our technological achievement. However, our technological achievement, once seen as our greatest legacy has become the cancer eating away at humanity and the delicate ecological processes of Earth. Our technology has been driving us towards material comfort and gain in the consciousness of the "People of Maize," and has become the thing that will ultimately force our evolutionary leap.

As McKenna once stated to an audience of rapt listeners, "We are all gathered at the end game of developmental processes on this planet. We are all about to become unrecognizable to ourselves as a species. Our technologies, religions and science have pushed us towards this for years. We feel the tug of the transcendental and transformative."[5]

---

[5] Terrance McKenna Speech at University – date unknown

Time the Harbinger of the Transformative

"Do not dwell in the past, do not dream of the future, concentrate the mind on the present moment."

Buddha

As I pondered McKenna's insightful theories of time and its acceleration towards an event horizon, I asked myself how would this be possible? Doesn't this defy the laws of physics and in fact is an impossibility? Surprisingly the answer I've discovered is quite simple. Time is based upon human perception, thus time only exists within the mind of the being, in this case humans, who perceive it. And, like the physics of energy applied to a pendulum, as the object is furthest from the center its movement is slower, however as it progresses closer to the center it speeds up exponentially.

This is reflective of our physical and spiritual perception of history, and the events happening around us. If you are reading this book, I'm certain you feel the tug of God's creative energy pulling you with greater speed towards some 'event'. I'm certain this is reflected in your everyday life, were the days seem to blend together and flash before your eyes. This is the essence of what compelled me to write this book, and share my vision (truths) of humanities past, present and future.

In the billions of years it took for the universe to form, the slow evolutionary process played out over eons in time. However with the emergence of the conscious living organism slowly perceiving time, the events developed and transpired with greater speed reflective of the evolution of the consciousness inherent in the species. It is as if time (history) and light (energy) are woven together in a Divine Intelligent Matrix, waltzing with infinite consciousness – God.

This macrocosm of greater evolutionary processes, mirrors the microcosm of the evolutionary matrix of an ecosystem. As I have pondered, time only exists by the consciousness that perceives it and the events experienced within it.

With this in mind, could the very nature of time, and the cosmic cycles of evolution be the 'thing' that holds the key to our next stage of evolution? Is this the unveiling spoken of by the ancients and prophets? The very nature of time, as we approach the present, appears to be more energetic in its surge towards change.

As science tells us, the early universe was pure plasma void of complex life. As the third dimensional universe expanded the complexity and diversity of life expanded and thus time had begun to speed up. The concept of the connectivity of time is connected to the evolutionary complexity of life. The universe by intelligent design is connecting-the-dots of manifestation in this physical dimension and leading humanity and in fact all life on Earth, and Earth itself, towards a quantum evolutionary leap.

And, as I've previously proposed, the universe does this through consciousness and its co-creative capabilities within the divine matrix of intelligent design. We can think of it as "Mental Data" that is brought into relationship with physical matter. The universal collective mind – the Christ Consciousness, under the divine guidance of God, is coordinating a collective view and in doing so becoming more intelligent and aware. It is this Christ

Consciousness that now flows freely on earth in the form of an unseen energy, some may call the Second Coming of Christ, that is awakening our souls from their dormant state to take part in the Round Dance of the Cross, and at long last sup with God our creator.

The very evolutionary process of humans illustrates the point I and other scholars, such as McKenna have pondered, which is the distinctly unusual process by which we find ourselves in our evolutionary path. Science still wants to hold firm to the argument that humans evolved from primates, but somehow miraculously achieved this evolution from primate structures to Homo Sapiens within a very short period of time in Earth's evolutionary process within the animal kingdom.

Unlike animals, humans haven't seemingly evolved physically since the introduction of anatomically modern man some 200,000 years ago. Ask yourself, why there was a sudden and dramatic leap in the physical evolution from a primate, only to remain stagnate anatomically since that time. Could it be that the intelligent design, the divine will of God, chose human evolution to be transformative within its consciousness rather than within its DNA? Could it be that our DNA is already encoded to evolve mirroring the physical and spiritual cycles?

Modern man's evolution is seen in our epigenetic awareness of ourselves expressed in our language, art, music, technology and societal structures. This physical manifestation of our collective conscious is what is driving our evolutionary change. This epigenetic phenomenon is intertwined with time to race us exponentially towards the quantum metamorphosis that awaits us.

In speaking of time Terrance McKenna said, "every day is composed of four other days, and time is a resonance created by other time … but somehow a day centuries ago in the past meets in the now to create a unique moment or event in current time."

Meaning, is history repeating itself as we race towards the inevitable event? In the year 2012, as we approach the Point of Singularity, is humanity living in both the present and past – repeating the rise and fall of civilizations of eons ago simultaneously? Is this what the Mayans referred to as "..all will be revealed and we will remember who we are"?

I believe we are reliving our past lives, simultaneously in the present, so that we may transcend into our future with the blessings of God in this galactic return to source energy. This would certainly explain the unfamiliar energies that seem to permeate the very air we breathe. And in my case, it helps me to understand the unfamiliar and profound epiphanies, or what I simply call a 'knowing' that suddenly and with great strength move into my being to materialize with clarity in my mind and consciousness.

In pondering the very nature of time, I've heard the analogy of a sand dune used to illustrate this point. If you are to visualize sand dunes you will often notice that they look like wind – the sand dune forms the shape of the wind. Well, you may ask, what is wind? As science explains, wind is a pressure variant phenomenon that fluctuates over time. In a way the sand dunes that look like the wind are metaphorically a lower variant of wind. Thus, the shape of which a sand dune is given by wind is a physical example of a lower dimensional image of time.

Now, instead of sand dunes think of genes. Our genes can be understood as a lower dimensional reflection, or imprint, of the higher dimensional force that created it. As I propose, this omnipotent force (God) that created matter left an imprint of itself within all genetic makeup in the material dimension, much like the image of a sand dune reflecting the wind that shaped it. Similarly our material world is but an illusion in a finite existence, and our soul in its eternal, infinite existence imbues the same imprint of God's consciousness. We are physically and spiritually a cosmic dream in the mind of God, gifted with its fragmented essence and eternal life.

In the physical world, humans understand that it is our 'humanity', our ability to reason through mental faculty that sets us apart from the rest of the animal kingdom. We shouldn't view ourselves as better than the lesser conscious of animals for all life is imbued with the same 'imprint' from our creator and thus divine and eternal in its own way.

Ours is the higher gift of reason coupled with the existence of an ever evolving eternal, infinite consciousness as the Sons and Daughters of Light.

Wisdom of the Ancients

"O men, list to the voice of wisdom: list to the voice of light. Mysteries there are in the cosmos that unveiled fill the world with their light. Let he who would be free from the bonds of darkness first divine the material from the immortal, the fire from the Earth; for know ye that as Earth descends to Earth, so also fire ascends unto fire and becomes one with fire. He who knows that the fire [light] that is within himself shall ascend unto eternal fire [light] and dwell in it eternally."

Thoth

Decoding our past is a key ingredient to understand our future. The allegories and metaphors of ancient civilizations are the breadcrumbs leading humanity down its greater truth towards this point of singularity and the awaiting evolutionary leap of consciousness. It is both a physical and spiritual rebirth in the true light of God, the benevolent creator above everything.

The stories told in the three-main monotheistic religions of Judaism, Christianity and Islam have sadly confused humanity to the point of hatred and frustration in the need to believe that their chosen dogma is the only path to God. However, the true path towards the realization of the Golden Age is simple, and already imbedded within the cycles of time and our DNA.

To be "saved" is simply to awaken to the truth. This is the simple message bestowed by Jesus, a truth that was hidden from humanity for centuries' as it did not fit the political agenda of those who sought to use religion to control the masses. Concerning the nature of existence, Jesus is quoted as saying in the Gospels of Thomas, *"If they say to you, 'Where do you come from?' Say, 'We come from the light; the place where the light [first] came into being . . .' If they say to you, 'Who are you?' Say, 'We are the children of the light, and we are the chosen of the living Father.' If they ask you, 'What is the sign of your Father in you?' Say to them, 'Movement and rest.'"*[6]

This cosmic cyclical event ushering humanity towards singularity is also forcing us to re-examine our history to its very beginning, and look to the ancient texts and the stars for guidance and wisdom. Reason and science can't explain the event that lies before us. Many believe this evolution of consciousness, many call the "ascension", and spiritual rebirth can be understood in the legends of ancient demi-gods, who may have in fact been flesh and blood humans who once walked the Earth thousands of years ago. Over time, their stories became the myths and legends of the Gods of Egypt, Maya, Greek, and Rome all bearing a strikingly similar story, but with different names.

A pivotal persona of such an ancient deity is the Egyptian Goddess Isis, who throughout the ages can be seen as the Virgin Queen representative of the Divine Feminine. This is the iconic image illustrated by the Virgin Mary, and, yes, another extraordinary woman who personified the Divine Feminine, Mary Magdalene, whose life we will discuss in this book.

An inscription of Isis written thousands of years ago that mirrors passages in Biblical texts states: "I, Isis, am all that has been, that is or shall be, no mortal Man hath ever unveiled."[7]

The "I AM" being a central point mentioned on numerous

---

[6] The Gospel of Thomas
[7] The Secret Teachings of All Ages, 1928, Manly P. Hall

occasions by Jesus. The concept of the "I AM" was known and followed by the ancients long before the birth of Jesus, and it was Jesus who lovingly returned this truth to humanity.

The "I AM" power is the healing power of the universe. It's the concept by which the universe and all life are made. The "I AM" power can be experienced as a light or as a profound feeling in the heart or as great wisdom. There is no limit to the power of the "I AM" to heal and transform every aspect of life. Jesus and Ascended Masters knew the healing and life-transforming power of the "I AM" and sought to teach humanity this truth. The ancients knew that using these words is a call to the universe, to the healing power, to God, asking that the doors be opened to release pure and radiant energy into our life.

This ancient Egyptian deity, Isis – saying of herself "I AM", is the Virgin Goddess who gave birth to all living things chief among them is the sun. Isis worship in its essence is worship of our Sun. She is the Queen of Heaven – the Cosmic Queen. Isis has an incredible story that surprisingly pre-dates even the ancient Egyptian civilization.

According to Augustus Le Plongeon (a famous 19th Century historian and photographer who studied pre-Columbian ruins and the ancient cultures of Central America, and was the earliest known proponent of Mayanism), "the story of Isis can first be seen in the mythical story of a woman [a high priestess] who once existed in Central America amongst an ancient civilization called the Mayans."[8] Le Plongeon believes that the Mayans of Central America (i.e. the highlands of Guatemala and the Mexican Yucatan) pre-date the Egyptian civilization.

To explore this amazing theory we must open the door to a long ago and forgotten advanced civilization that met its demise in a cataclysmic flood and earthquake. If you haven't already guessed,

---

[8] The Secret Teachings of All Ages, 1928, Manly P. Hall

I'm speaking about the lost civilization of Atlantis. In recent times, and quite possibly due to the quickening of time towards singularity, historical texts and references have emerged with great clarity and strength affirming the existence of a highly advanced human civilization called Atlantis. It is Plato who speaks about this extraordinary civilization, in which the Egyptians and Greeks modeled their deities after.

Within these texts are numerous references, which suggest that prior to the cataclysmic flood, God selected an enlightened few who followed the truth and light, to be the torchbearers imparting wisdom and knowledge to future generations. It is the survivors of Atlantis who managed to flee to Central America, Egypt and Tibet prior to the final cataclysmic sinking of their continent.

Atlantis was believed to be a kingdom more connected to nature and earth, harnessing solar power and crystals creating cities of light built from natural stone. The pyramid is a critical component of their spiritual and mystical beliefs, as this structure was designed to harness the divine energies of the heavens, and inter-dimensional spaces to Earth for their enlightened state of existence. In the lore of Atlantis, the myth of their destruction is shrouded in an 'elite' class of Atlanteans who sought personal material gain with such fever that they induced black magic, conjuring up dark forces to achieve their corrupt aim. If you haven't read my previous book, *Smoke and Mirrors*, I recommend you get a free copy on my website. www.smokeandmirrorsbook.com

The similar force that is at play in our world today is revealed in this book.

Within this priestly structure in Atlantis towards its last days, there existed a Dark Brotherhood and a White Brotherhood. One brotherhood clearly represents the 'unconscious' dark forces of human nature – the 'antichrist', and the other represents the 'awakened – Christ Consciousness'. Some members of the White

Brotherhood, called the Nakkal, were chosen by God to survive so that darkness doesn't completely overtake future civilizations, and that humanity can be forewarned about the cosmic cycles of time. [9]

Chief among the Nakkal was the bright star, the ascended master, Thoth the Atlantean, who with his wife, Seshati, would become the builders of the Great Pyramid of Giza and Sphinx in Egypt, and impart the Emerald Tablets to humanity.

Thoth (also known as Hermes Tresmigistus, and Mercury founder of the Hermetic tradition) became a deity in later generations of Egyptians, Greeks and Romans viewed as Hermes the God of Wisdom. Thoth's wife, Seshati, would be remembered in Ancient Egypt as the Goddess of knowledge and writing. Seshati (also spelled Safkhet or Safekh – meaning "Seven") is the scribe and record keeper with her name meaning 'she who scrivens or she who is a scribe'. Seshati is credited with inventing writing, as well as architecture, astronomy, astrology and building mathematics. She is depicted with seven-pointed flower pedals above her head, signifying the divine energies derived from opening the seven chakras in the human body and the divine kundalini energy connection to God.

Seshati was the Atlantean who performed the "stretch the cord" ritual to heaven and ensured the pyramids were laid out to the correct measurements and their dimensions were precise to ensure the divine alignments necessary to serve as a structures, which would open the divine portals to heaven and track the cosmic cycles – the Round Dance of the Cross. Although some historian's who have studied the mythology of Atlantis, give Thoth credit as the builder of the Great Pyramid of Giza, it may be that his wife, Seshati was its architect and inspiration.

In the final demise of Atlantis some 13,000 years ago, not surprisingly coinciding with the last, most recent Round Dance of the

---

[9] The Serpent of Light: Beyond 2012, 2007, Drunvalo Melchizedek

Cross, our ancient history does seem to be repeating in our present moment. However, unlike the demise of Atlantis, our choice within our collective consciousness is yet to be made.

A crowning achievement of Ancient Egypt is the Great Pyramid of Giza, falsely accredited to the Pharaoh Cheops who never asserted it was his creation. The Atlantean culture centered on the natural environment, cosmic cycles and inter-dimensional energies, and they understood the importance of leaving pyramids along the lay lines of earth that would serve as markers for future civilizations to re-learn the truth of humanity's history on earth, our divine origin, and the cosmic cycles which govern our evolution.

These pyramids can be seen from Central America, to North Africa to Asia, and the Atlantean civilization theory when applied to our current understanding of events explains what has previously been, for so long, the unexplainable. As mythology and such notable scholars as Plato suggest, these Atlanteans travelled the world building their pyramids and teaching more primitive humans of the truth of their divine creation.

Returning back to the incredible genesis of the Goddess Isis, Le Plongeon states, "In ancient Mayan culture Isis was called the Goddess Moo. After the death of her husband, Prince Coh – similar to Osiris, Queen Moo was forced to escape and sought refuge amongst the Mayan colonies of ancient Egypt."

These Mayan-Egyptians, the Atlantean White Brotherhood Nakkal, re-named this Cosmic Queen, Isis, and thus began her legend.

It is believed these Atlantean colonies pre-date 10,000 BC and their civilization may have existed as far back as 65,000 BC, the equivalent of two and one-half 26,000-year cycles. Thoth, the Atlantean, speaks about the events of the last days of Atlantis, his spiritual ascension to become an Ascended Master, the nature of God and cosmic cycles, and the importance of the Great Pyramid of Giza at great length in his Emerald Tablets. I recommend you read *"The Emerald Tablets of Thoth the Atlantean"* by M. Doreal. To give you a better understanding of this miraculous story, from Emerald Tablet I, Thoth says of his beloved Atlantis and Great Pyramid of Giza:

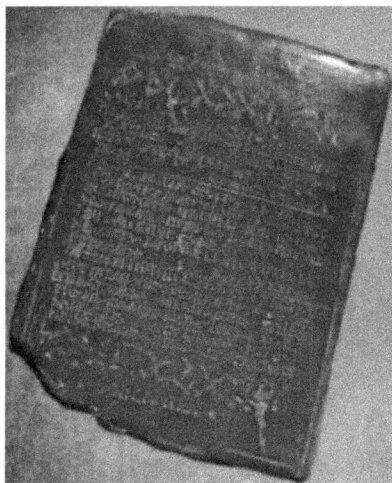

"I, THOTH, the Atlantean, master of mysteries, keeper of records, mighty king, magician, living from generation to generation, being about to pass into the halls of Amenti, set down for the guidance of those that are to come after, these records of the mighty wisdom of Great Atlantis.

In the great city of KEOR on the island of UNDAL, in a time far past, I began this incarnation. Not as the little men of the present age did the mighty ones of Atlantis live and die, but rather from aeon to aeon did they renew their life in the Halls of Amenti where the river of life flows eternally onward.

A hundred times ten have I descended the dark way that led into light, and as many times have I ascended from the darkness into the light my strength and power renewed. Now for a time I descend, and the men of KHEM [Khem is ancient Egypt] shall know me no more. ..

Great were my people in the ancient days, great beyond the conception of the little people now around me; knowing the wisdom of old, seeking far within the heart of infinity knowledge that belonged to Earth's youth.

Wise were we with the wisdom of the Children of Light who dwelt among us. Strong were we with the power drawn from the eternal fire.

And of all these, greatest among the children of men was my father, THOTME, keeper of the great Temple, link between the Children of Light who dwelt within the temple and the races of men who inhabited the ten islands.

Mouthpiece, after the Three, of the Dweller of UNAL, speaking to the Kings with the voice that must be obeyed.

Grew I there from a child into manhood, being taught by my father the elder mysteries, until in time there grew within the fire of wisdom, until it burst into a consuming flame.

Naught desired I but the attainment of wisdom. Until on a great day the command came from the Dweller of the Temple that I be brought before him. Few there were among the children of men who had looked upon that mighty face and lived, for not as the sons of men are the Children of Light when they are not incarnate in a physical body.

Chosen was I from the sons of men, taught by the Dweller [God] so that his purposes might be fulfilled, purposes yet unborn in the womb of time.

Long ages I dwelt in the Temple, learning ever and yet ever more wisdom, until I, too, approached the light emitted from the great fire.

Taught me he, the path to Amenti, the underworld [heaven] where the great king sits upon his throne of might. Deep I bowed in homage before the Lords of Life and the Lords of Death, receiving as my gift the Key of Life.

Free was I of the Halls of Amenti, bound not by death to the circle of life. Far to the stars I journeyed until space and time became as naught.

Then having drunk deep of the cup of wisdom, I looked into the hearts of men and there found I greater mysteries and was glad. For only in the Search for Truth could my Soul be stilled and the flame within be quenched.

Down through the ages I lived, seeing those around me taste of the cup of death and return again in the light of life.

Gradually from the Kingdoms of Atlantis passed waves of consciousness that had been one with me, only to be replaced by spawn of a lower star.

In obedience to the law, the word of the Master grew into flower. Downward into the darkness turned the thoughts of the Atlanteans, until at last in this wrath arose from his AGWANTI, [this word has no English equivalent; it means a state of detachment – the 'unconscious' – the 'antichrist'] speaking The Word, calling the power.

Deep in Earth's heart, the sons of Amenti heard, and hearing, directing the changing of the flower of fire that burns eternally, changing and shifting, using the LOGOS [the Word of God], until that great fire changed its direction.

Over the world then broke the great waters, drowning and sinking, changing Earth's balance until only the Temple of Light was left standing on the great mountain on UNDAL still rising out of the water; some there were who were living, saved from the rush of the fountains.

Called to me then the Master, saying: Gather ye together my people. Take them by the arts ye have learned of far across the waters, until ye reach the land of the hairy barbarians, dwelling in caves of the desert. Follow there the plan that ye know of.

Gathered I then my people and entered the great ship of the Master. Upward we rose into the morning. Dark beneath us lay the Temple. Suddenly over it rose the waters. Vanished from Earth, until the time appointed, was the great Temple.

Fast we fled toward the sun of the morning, until beneath us lay the land of the children of KHEM [Egypt]. Raging, they came with cudgels and spears, lifted in anger seeking to slay and utterly destroy the Sons of Atlantis.

Then raised I my staff and directed a ray of vibration, striking them still in their tracks as fragments of stone of the mountain.

Then spoke I to them in words calm and peaceful, telling them of the might of Atlantis, saying we were children of the Sun and its messengers. Cowed I them by my display of magic-science, until at my feet they groveled, when I released them.

Long dwelt we in the land of KHEM [Egypt], long and yet long again. Until obeying the commands of the Master [God], who while sleeping yet lives eternally, I sent from me the Sons of Atlantis, sent them in many directions, that from the womb of time wisdom might rise again in her children.

Long time dwelt I in the land of KHEM, doing great works by the wisdom within me. Upward grew into the light of knowledge the children of KHEM, watered by the rains of my wisdom.

Blasted I then a path to Amenti so that I might retain my powers, living from age to age a Sun of Atlantis, keeping the wisdom, preserving the records. Great grew the sons of KHEM, conquering the people around them, growing slowly upwards in Soul force.

Now for a time I go from among them into the dark halls of Amenti, deep in the halls of the Earth, before the Lords of the powers, face to face once again with the Dweller. Raised I high over the entrance, a doorway, a gateway leading down to Amenti.

Few there would be with courage to dare it, few pass the portal to dark Amenti. Raised over the passage, I, a mighty pyramid [The Great Pyramid of Giza], using the power that overcomes Earth force (gravity). Deep and yet deeper place I a force-house or chamber; from it carved I a circular passage reaching almost to the great summit.

There in the apex, set I the crystal, sending the ray into the "Time-Space," drawing the force from out of the ether [the energy that connects dimensions – the unseen mental dimension], concentrating upon the gateway to Amenti. Other chambers I built and left vacant to all seeming, yet hidden within them are the keys to Amenti. He who in courage would dare the dark realms, let him be purified first by long fasting.

Lie in the sarcophagus of stone in my chamber [now the Ascended Queen's Chamber of the Great Pyramid of Giza]. Then reveal I to him the great mysteries. Soon shall he follow to where I shall meet him, even in the darkness of Earth shall I meet him, I, Thoth, Lord of Wisdom, meet him and hold him and dwell with him always.

Builded I the Great Pyramid, patterned after the pyramid of Earth force, burning eternally so that it, too, might remain through the ages.

In it, I builded my knowledge of "Magic-Science" so that I might be here when again I return from Amenti, Aye, while I sleep in the Halls of Amenti, my Soul roaming free will incarnate, dwell among men in this form or another. [Reincarnating as Hermes Tresgimistus, and subsequent Ascended Masters who have walked on Earth.]

Emissary on Earth am I of the Dweller [God], fulfilling his commands so many might be lifted. Now return I to the halls of Amenti, leaving behind me some of my wisdom. Preserve ye and keep ye the command of the Dweller: Lift ever upwards your eyes toward the light.

Surely in time, ye are one with the Master [God], surely by right ye are one with the Master, surely by right yet are one with the ALL. Now, I depart from ye. Know my commandments, keep them and be them, and I will be with you, helping and guiding you into the Light. Now before me opens the portal. Go I down in the darkness of night."

As described by Thoth, I've been in the ascending Queen's Chamber in the Great Pyramid of Giza and meditated in front of and inside the sarcophagus – on October 12, 2012, just as Thoth described.

Yes, the transcendental experience to wisdom and truth awaits the student who with an open heart journeys to meet Thoth.

In unraveling this ancient history, I believe the true story of the Atlantean Thoth, and his wife Seshati, can be seen in the metaphorical stories of the Mayan Queen Moo and Prince Coh, who right or wrong, Le Plongeon attributes to the myth of Isis and Osiris. It appears that these ancient demi-Gods were in fact the legends of people who once existed on Earth, and their story and the messages that they left behind may be one of the central keys to realizing the Christ Consciousness and the Ascension in this our 26,000-year mark – the third full galactic transit since the beginning of Atlantis.

The Ascension

*Jesus said, "There is light within a person of
light, and it lights up the whole universe. If it
does not shine, there is darkness."*
                                        The Gospel of Thomas

Reincarnation

"The souls must re-enter the absolute substance whence they have emerged. But to accomplish this, they must develop all the perfections, the germ of which is planted in them; and if they have not fulfilled this condition during one life, they must commence another, a third, and so forth, until they have acquired the condition which fits them for reunion with God."

Zohar, one of the principal Cabalistic texts

Buddhism and Hinduism are two well-known religions still teaching the concept of reincarnation, however what is unknown to most people today is that the same belief was taught by the beloved teacher of light, Jesus the Christ. Reincarnation and the concept of a soul living numerous lives on Earth in this third dimension is nothing new and in fact is as old as recorded human history. Sadly this truth has succumbed to history and the obsession of the material illusion. The pre-existence of the soul was a teaching held by early Christians until it was condemned by the Roman Catholic Church in 553 A.D., as it challenged their orthodox doctrine. Despite the efforts of the Roman Catholic Church, this belief resided amongst Christians until the 10th Century, until it met its demise in the annals of history.

The very idea, that religious dogma implies that when a person dies their soul would sleep in the grave along with their corpse waiting for the "last day and the final judgment" is representative of a cruel God. This certainly is not the God that I know and love deeply. This is not the God that Jesus, himself, and the many prophets who have walked on Earth spoke of.

The logical question is why has this truth been kept a mystery in the Christian faith. Like so many beliefs, it became victim to politics and control. In the case of Christianity, the Roman faction rejected the concept of reincarnation, although this concept was spoken of by Jesus, himself. The Roman orthodox assertion is that Jesus was in fact God, a concept that Jesus repeatedly denied to his followers. The truth is that Jesus never suggested he was God, but rather the "Son of Humanity" showing 'the truth, the way'.

Jesus was once a living example of the human-divine that exists in all of us in the awakened state of Christ Consciousness. His message is quite simple, it should be the aim of all souls to look past the illusion of the material world and join with God in the sea of the divine consciousness, thus defeating the need for reincarnation. Sadly for the last 1,500 years, Rome won the political battle to perpetuate their orthodox concept of the resurrection, the so-called "Night of the Living Dead". As we approach our moment of truth and singularity, God frees us from these oppressive ideas to let our souls soar to new heights.

In this, we are reminded of the simple story of a professor who went to visit with a famous Zen master – a Buddhist master. While the master quietly served tea, the professor talked about Zen. The master poured the visitor's cup to the brim, and then kept pouring. The professor watched the overflowing cup until he could no longer restrain himself. "It's overfull! No more will go in!" the professor blurted. "You are like this cup," the master replied, "How can I show you Zen unless you first empty your cup." How this simple metaphor resonates within our soul in this present moment.

With even simple investigation, through the eyes of "an empty cup", you can learn of the many references in religious texts pointing to the truth of reincarnation. However, I want to point your attention to a man who was not only one of the closest Apostles of Jesus, but was his half-brother and the twin of Mary Magdalene. I'm speaking about Thomas Didymus, and his gospels, which many truth seekers view as imbuing the essence of the divine message – 'the truth, the way'.

As you read some passages from the Gospel of Thomas that I present for your consideration, you will discover the amazing message Jesus sought to convey and the simple truth about reincarnation. The story of both Thomas Didymus and Mary Magdalene has a special and personal meaning to me, and I am deeply honored to share this truth with you. I hope that the true message they conveyed, now re-discovered, will ignite the light in your soul so that you too can follow the breadcrumbs to God.

Thomas was among Jesus's most devoted disciples and carried a realization, which may not have been shared within any of the other gospels except for his twin sister, Mary Magdalene. This realization, which he imparted to humanity in his gospel, is the truth about mankind's potential, why we struggle during our lives, and of course the truth of the cosmic "drama" of reincarnation. As we explore the very idea that we approach a singularity, understanding the nature of our soul and releasing judgment is the Rosetta stone to decode our destiny.

Perhaps a most profound passage which illustrates Jesus' belief in reincarnation and the importance of reconciling your past with your present is seen in this passage from the Gospel of Thomas, *"Jesus said, 'When you see your likeness [in a mirror] you are pleased; but when you see your **images**, which have come **into being before you**, how much will you have to bear'."*[10]

This is a reference to the path to attain enlightenment. As many Ascended Masters have taught, a soul must reconcile its past lives, karmic forces, in order to ascend in higher consciousness. And,

---

[10] The Gospel of Thomas

as Thomas adds, *"Jesus said, 'Let the one who seeks not stop seeking until he finds. When he finds, he will become troubled. When he becomes troubled, he will be astonished."*[11]

Meaning the revelation of your self in its infinite existence over many lives will lead to astonishment – amazement.

According to Thomas, Jesus tells us that the Kingdom of God is inside each one of us, and is reflected in the miracle of life, as Jesus says, *"When you come to know yourself, then you will be known, and you will see that it is you who are the children of the living Father. But if you will not know yourselves, you dwell in poverty, and it is you who are that poverty."*[12]

As conveyed by Thomas, Jesus declared that we must find out where we came from, and go back and take our place "in the beginning." Jesus says, *"Blessed is the one who came into being before he came into being."*[12]

A final passage I want to share with you from the Gospel of Thomas that imbues the very essence of what I write about, *Jesus said, "If you bring forth what is within you, what you bring forth will save you. If you do not bring forth what is within you, what you do not bring forth will destroy you. "*[12]

Clearly we are being invited to embark on the journey of our many lives, and in doing so to surrender judgment of ourselves, doubt and fear of the unknown. We are being invited to accept this cosmic moment of truth, and re-discover God at the Round Dance of the Cross. It was only a short time ago, that I was propelled down this path, by the grace of God, and I was shown my many reflections in a mirror. In taking this journey, with an open heart, the abundance of universal wisdom has filled my very being. I metaphorically emptied my cup and let God fill it back up.

---

[11] The Gospel of Thomas
[12] Ibid

The Forgotten Story

Jesus said onto his Apostle's, of Mary Magdalene his Beloved Companion, "Leave her be. She has anointed me for what I am come to do, and done what she is appointed to do. Only from the truth I tell you, whenever they speak of me, what she has done will also be told in memory of her. You do not know or understand what she has done. I tell you this: When all have abandoned me, only she shall stand beside me like a tower. A tower built on a high hill and fortified cannot fall, nor can it be hidden. From this day forth, she shall be known as Midgalah, for she shall be as a tower to my flock, and the time will soon come when her tower shall stand alone by mine."

The Gospel of Mary Magdalene 32:4

It is difficult to convey the depths of truth I've been shown about the woman who was Mary Magdalene. Her story transcends time and the many Round Dances of the Cross. She embodies the Divine Feminine, and her earlier incarnations and legacy can be seen on Earth as far back as the last days of Atlantis. She was the face inspiring such iconic symbols as the Sphinx in Egypt, and the Statue of Liberty in the United States. She was a torchbearer lighting the way to the divine, the tower that Jesus spoke of.

Just over 2,000 years ago, Mary Magdalene would enter this dimension as a child born to Queen Cleopatra and Marc Antony. Her given name was Cleopatra Selene. Her twin brother would be remembered in history as Thomas Didymus Judas (meaning twin), and her younger brother as James the younger – noted in the Bible as a brother of Jesus. Thomas was born Alexander of Helios, and James was born Ptolemy Philadelphus. Their epic story with their half brother Jesus, the Christ was lost to the manipulation of history and mistranslations, but never the less the message that they conveyed as the closest Apostle's of Jesus is remembered in her story – the Gospel of Mary Magdalene.

Jesus, originally born Caesarion, was the child of Ptolemy Queen Cleopatra and Roman Emperor Julius Caesar. This great and divine soul entered our world as both the "Son of God", befitting his status as a Pharaoh of Egypt, and "King of Kings", befitting his status as the future Emperor of Rome and only son of Julius Caesar. He was remembered by his disciples as the "Son of Humanity".

Most people are well acquainted with the story of Jesus' anger directed at the corrupt moneychangers in the temples, but what most people do not know is that this is also the legacy of his biological father – Julius Caesar. It was Caesar who tried to break the monopoly on the usury of their Denarius coinage and limit the interest, which was bankrupting Rome and its citizens. Julius Caesar challenged the corrupt financial oligarchy of the Roman elites, and for this I argue, his life was taken in the assassination carried out by Gaius Cassius Longinus, and Junius Brutus.

Similarly, it was not until Jesus challenged the corrupt moneychangers that his fate leading to his crucifixion was sealed. It was within a few days of challenging the moneychangers that Jesus was crucified.

As said stated by author F.R. Burch in his book *Money and Its True Function*, "As long as Christ confined his teaching to the realm of morality and righteousness,

he was undisturbed: it was not till he assailed the established economic system and 'cast out' the profiteers
and 'overthrew the tables of the money changers', that
he was doomed. The following day he was questioned,
betrayed on the second, tried on the third and on the
fourth crucified."[13]

As legend tells us, it was the Spear of Longinus that pierced the
heart of Jesus on the cross, thus taking his life. No explanation has
ever been given as to why this spear would bear the same name as
Julius Caesar's assassin. However, I believe the answer is quite obvious. Jesus was Caesar, meaning the Emperor of Rome and legal
heir to the throne.

Another truth that has been hidden for centuries is the date of
Jesus's true birth, which I know to be December 25, 47 BC. Many
have argued that Jesus was given the birth date of December 25th
in honor of the Pagan celebration of the winter solstice. However,
what people do not realize is that it was his biological father, Julius
Caesar, who deemed December 25th the day celebrating the winter solstice in the year 46 BC – one year after the birth of his son
Caesarion, the "Son of God" and "King of Kings". Thus, the beloved date and celebration on December 25th does truly mark the
day of the birth of Jesus the Christ. What a beautiful affirmation
denied to humanity for the last 2,000 years.

After the death of Julius Caesar, Queen Cleopatra became inti-
mate with his close friend, Marc Antony. This relationship resulted
in the birth of three children who would go on to support Jesus in
his Ministry some 17 years later.

It was at the Donations of Alexandria ceremony, when
Caesarion was 13 years old and Selene and her twin Alexander

---

[13] The American Mercury, 1964, vol. 96-101, p. 18 from No More National Debt,
Bill Still, 2011, p.67

were 6 years old, Queen Cleopatra declared Caesarion the "Son of God" in honor of the infant son Horus born to the Virgin Queen Isis. Cleopatra then granted Caesarion the Kingdom of Egypt in his coming of age celebration. She then granted the lands east of Egypt, in a place called Cyrenaica (modern day Libya) to her daughter Selene, who was Caesarion (Jesus) half-sister. In that moment Selene became the Queen of Cyrenaica.

Once this announcement found its way to the corridors of Rome, Emperor Augustus (Gaius Octavius Caepias), Julius Caesar's adopted son who claimed the throne of Caesar following his death, felt threatened by Caesar's true son Caesarion. Augustus believed himself to be the Son of God, and publicly declared himself in Rome as "Filius Dei", Latin for "Son of God". Therefore, Augustus made it his mission to kill his greatest threat, Caesarion.

Queen Cleopatra being forewarned of this threat sent Caesarion into hiding with his beloved and trusted nanny, Mary of Bethany (real name Mary of Bithynia) and her husband Joseph. In the quiet hours of the night, the group left Egypt en route for India and Tibet. To further protect Caesarion, Mary who had raised this extraordinary child knowing he was sent to Earth by God for a divine purpose, re-named him a Hebrew name – Yeshua (also called Esa in Arabic and Jesus in English).

History books tell us that Caesarion was killed en route to India, however the truth is he escaped his would be assassins. It was his tutor, Rhodon, who falsely declared Caesarion was murdered to protect this divine young man from the Roman legions and Emperor August's personal mercenary King Herod, who was named by Augustus to be the Procurator of Syria.

Upon arrival to Tibet, where Jesus spent the 'missing years' of his life he was welcomed by the Buddhist monks who saw in him the Son of Humanity, and the divine nature of his soul. The holy monks declared him the "Buddha Issa", meaning the Ascended

Master and teacher of God's kingdom of light – Jesus the Christ.[14]

Sadly, Queen Cleopatra and Marc Antony committed suicide rather than be taken by Augustus' Roman legions, and thus they left Selene and her brothers to the mercy of the Roman Empire. Selene and her brother's legacy would disappear in the forgotten records of time, and the truth of who they were in Jesus' ministry would be forgotten in mistranslations and within the political and social manipulation of the New Testament and the religion Christianity.

This is a brief glimpse of the beautiful story that came to me in a series of events, visions, dreams and subsequent research. During one of my many trips to Libya in May 2012, after publishing my first book *Smoke and Mirrors* where I detail this story and provide the historical context to support it, I had the opportunity to meet a Libyan General in the new government. This General is unlike any General I have met in North Africa. He is truly a kind man, with kind eyes and a good heart who has a deep love for Jesus as well, even naming one of his son's Esa (Jesus).

One evening over coffee, I told the General the story of Cleopatra Selene and her connection to Libya as the Queen of Cyrenaica. The General immediately picked up his phone and made a few calls that would propel me on a miraculous adventure. The following day I boarded a plane from Tripoli for Benghazi, and drove 3-hours east to Cyrene (modern town of Shahat Libya). Cyrene is the site of ancient Roman ruins that amongst the many spectacular things to see, marks the birthplace of Apostle Mark, who would be one of two Apostles who identified the Marys who stood at the foot of the cross.

---

[14] The Lost Years of Jesus: Documentary evidence of Jesus' 17-year journey to the East, 1984, Elizabeth Clare Prophet

It was Apostle Mark who in the Bible, 15:40 states, "There were also women looking on afar off: among them were Mary Magdalene and Mary the mother of James the younger."

Unfortunately misinterpretation and poor translation has led future generations to believe that there were three Mary's at the cross – the Virgin Mary, Mary Magdalene and Mary Cleophas – the mother of James the younger. However the truth is, Mary was the sister of James, with James being her younger brother, and Mary Cleophas was wrongly translated to imply the Virgin Mary's sister when a true translation in ancient Greek of Cleophas means "Cleopatra".[15]

Thus both Apostle Mark and Apostle John had described only two Mary's who stood at the cross of Jesus's crucifixion, the extraordinary and divinely ordained woman, the Virgin Mary, who would raise Jesus and be his true mother, and Jesus' beloved companion Mary Magdalene – his wife. Most biblical scholars do admit that the Apostle in the Bible, James the younger, was a brother to Jesus.

As I travelled to Cyrene I was searching for archeological evidence that Selene was Mary Magdalene, as God had revealed to me. For the last 42 years, Gaddafi had severely restricted foreigners movement to the ancient Roman ruins at Cyrene, but now with his

---

[15] Agape Bible Study of the New Testament

passing I was able to make the journey. And, as you may have already presumed, God did bless me with the evidence I was seeking.

In Cyrene there is a natural encampment of rocks within which has formed a spectacular natural pool off the Mediterranean, which is called "Cleopatra's Pool". Also in Libya, to the east, there is the "Villa of Selene", and of course the birthplace of both Apostle Mark and St. Simon of Cyrene. As the archeological evidence began to mount, I was given a precious gift and message from God.

As I passed by some Roman ruins not far from Cleopatra's Pool, a Libyan historian and archeologist, a professor whose name will remain anonymous, suddenly stopped. I, and others who were travelling with me, watched as he picked up what looked like a small pebble. I watched the professor, who I had just met that day, gently dust the dirt off and then smile and hand it proudly to me for inspection. It was an ancient Roman coin depicting simply "S". Below is a picture of this magnificent ancient coin.

Heart racing, I then asked the Professor what the "S" meant. He told me he did not know. Although he had found endless coins during his 30-years of excavation at the site, he had never seen such a coin. He wasn't even sure what the capital "S" meant.

Upon doing research the next day, God presented me with the answers I was seeking. The "S" coin, so beautiful and yet simple a symbol, has shown me a truth that goes back to the last days of Atlantis. It is the epic story of a Soul who had incarnated on Earth numerous times as the persona of **S**eshati from the time of Atlantis, the Queen of **S**heba from the days of King Solomon, and Cleopatra **S**elene (Mary Magdalene) the wife of Jesus, the Christ.

Such notable authors as Ralph Ellis and Dr. Robert Powell have put forward various pieces of the puzzle on the **"S"** in their published works. In the Hermetic tradition of the Ancient Egyptian Mystery schools, a tradition taught to Jesus who was inducted into these ancient mystery schools in Thoth's Great Pyramid of Giza, the "S" is the Alpha and the Omega, meaning the beginning and the end. As many scholars studying the depths of the Emerald Tablets and legend of Thoth and his wife Seshati believe, Thoth reincarnated on Earth as Hermes Trismegistus during a time before Moses was born in ancient Egypt. It would be Hermes who would impart the knowledge and wisdom of the Hermetic principles to mankind, thus the "S" of the Alpha and Omega.

Ralph Ellis in his book *Eden in Egypt*, points out another profound connection to the "S". The "S" is "Sefekh" the Egyptian number for "Seven".[16] As you may recall Seshati was also called Sefekh and was given the ancient symbol of Seven, to mark her divine authority.

As you will read in an excerpt I present from the Gospel of Mary Magdalene discovered in Egypt in 1896, Jesus called her "Magdal of Sefekh" meaning his Tower of Seven, which also translates to "Sheba" and is identified with the door to the Great Pyramid of Giza – Thoth and Seshati's architectural wonder and legacy of Atlantis. Jesus, the ascended master, whose consciousness resided in the sea of universal consciousness, was acknowledging her many lives and destiny by giving her this title.

> As Ralph Ellis states, "While it would appear that the mother of Jesus was following in these ancient Egyptian traditions, there is another section of the New-Testament evidence that again points towards Egypt... Mary Magdalene has been given some bad press in the Bible, and this was probably due to a strong priestly desire to distance her from Jesus – it was too embarrassing to admit that she was Jesus's wife, let alone contemplate the possibility that she was his sister too. But some of this bad press may actually have been derived from a subtle and deliberate mistranslation of a particular verse in the gospel of Luke: 'And certain women, which had been healed of evil spirits and in-

16 Eden in Egypt, November 25, 2008, Ralph Ellis

firmities, Mary called Magdalene, out of whom went seven devils." Luke 8:2[17]

However another more accurate translation of this same verse, Luke 8:2, reads, "And a certain woman who had worshipped blinding (strong) winds, Mary, named (after the) Seven Tower out of which went god (winds).[18]

This returns us to the broader theme of the book Approaching Singularity and the ascension to Christ Consciousness. It is the Divine Feminine embodied in the mythology of the Virgin Queen Isis, who walked on Earth as Seshati, Sheba and Selene (Mary Magdalene) that illustrates the ascension.

As Jesus indicates, *'the Queen of Sheba, the Tower of Seven, will stand in judgment on the human race'*. The veiled meaning behind these powerful words reflects a divine message from Jesus, which is Mary Magdalene's destiny.

Ralph Ellis proposes, "If the great entrance door to the Great Pyramid [of Giza] was unbolted and opened on the seventh day, and if the door itself was considered to be the 'Great Door of Heaven' – the divine portal that led into the nether world of the Djuat, the land of the gods and the dead – then it is axiomatic that Sefekh, the Egyptian term for unbolting a door and the number seven, would have become intimately associated with the Egyptian word Seba [Sheba], meaning the Great Door of Heaven."[19]

Many ancient cultures speak of this time as a transit to the Divine Feminine and the Golden Age, and it is through the Divine Feminine that the door to heaven is opened and the Soul will once again find its way to the Christ Consciousness. Jesus himself foretold of this destiny – the destiny of his wife, Mary Magdalene.

———•+•———

[17] Eden in Egypt, November 25, 2008, Ralph Ellis
[18] Ibid.
[19] Ibid

Mary Magdalene's Ascension Into Christ Consciousness

And Jesus answered him, "Only from the truth I tell you, unless you change and become as little children, you will not gain the Kingdom. When you make the Two into One, and when you make the Inner like the Outer and the Outer like the Inner, and Upper like the Lower, and when you make Male and Female into a single one, so that the male will not be male nor the female be female, when you make eyes in the place of an eye, a hand in place of a hand, a foot in place of a foot, an image in place of an image, only then shall you gain the kingdom."
The Gospel of Mary Magdalene 30:12

Although Kundalini is a term that comes from the Hindu tradition, it is a concept that was handed down to humanity thousands of years ago. Many Ascended Masters speak of this truth pre-dating even the high civilization of Atlantis. Kundalini refers to a powerful energy force (life-force) that resides in our body and connects us to the divine source of creation – God. People who are immersed in the illusion of the material world are completely unaware of the power within them, and thus this mysterious energy lies dormant in an un-awakened state.

Once this divine energy is awakened it will facilitate the higher spiritual connection to the divine, and the vast sea of Christ Consciousness and light will fill your being. In addition to the

Hindu tradition, Tibetian Buddhists also practice this powerful form of meditation, which they call "tummo" or "inner fire". Through the Kundalini awakening you will attain spiritual realization, profound wisdom and joy. In the Kundalini state you are basking in the light of God's kingdom while in human form and connecting to your divine soul, the very imprint of God.

The Kundalini is a form of meditation that focuses on key energy centers in the human body, called chakras. It is the release of these powerful energies that climb up the spinal cord to the crown of your head, and hence to the universal sea of consciousness. Each of the seven chakras is a mirror image to the seven dimensional planes of existence. As the energy rises up your spine, knowledge of each of the seven dimensions will flow through you. Thus achieving a divine enlightenment.

Attaining the Kundalini awakening is not so easy as relaxing the mind and meditating. In most cases it is a life long practice that must be coupled with the release of fear, judgment and doubt. Negative emotions, which we hold inside, block the Kundalini awakening. Whether you chose to harness your divine, innate power through the Kundalini or not, it's important that you forgive yourself and others, and that you view the world through the eyes of a child with wonder and love for all. This is the path and the way so many ascended masters like Jesus, the Christ, showed us.

I'm certain it will amaze and surprise you to learn that Jesus practiced and taught the Kundalini. This is the gift he gave to his wife, Mary Magdalene. Jesus showed her how to harness her innate divine power and thus free her soul from the material illusion. Through his light and love, Jesus sought to free us all.

I want to share with you a hidden secret, told within the Gospel of Mary Magdalene. As you read her story below, you will see the deep love and bond Jesus had with her, whom he called his "Beloved Companion". You will see that she was chosen by Jesus to lead the way, as the Tower of Seven. She was chosen as the human embodiment of the Divine Feminine and Christ Consciousness through the awakening of the seven chakras. It was her mission, after the passing of Jesus, to teach this as a part of Jesus' ministry and fulfill her destiny in the Round Dance of the Cross.

Sadly in the male-dominated world, Mary's truths were buried by jealous men seeking control, and advantage. In the most egregious manner they defamed Jesus and her legacy. Yet Jesus is truly a loving and forgiving soul, and thus forgives all souls knowing that we are like children. Now, in this moment of our Round Dance of the Cross the 'Second Coming of Christ' is upon us. This isn't the physical presence of Jesus, but rather the energy of God emanating from the center of our Milky Way galaxy. It is this creative energy that is flowing on Earth today, and crying out to humanity to awaken to our destiny and join the sea of cosmic consciousness.

From the Gospel of Mary Magdalene comes the story of the deep love between a husband and wife, and the gift of her ascension:

> Jesus says the following, "Only from the truth I tell you, there is one amongst you who has had my commandments, and keeps them. That One, is the one who loves me, and that one who loves me is also loved by me, and by the Spirit. To that one [she] I will reveal myself so that you will know that what I have said to you is true, that I am in Spirit as the Spirit is in me. And that same one [she] will the spirit complete in all ways, so that by this Sign you may know my words are true, and that my testimony is of the Spirit [God], the one who sent me." 35:17
>
> As Jesus continues, "Only from the truth I tell you, those amongst you who understand and keep my

commandments will not taste of death." 35:18 "The Beloved Companion [Mary Magdalene] followed Yeshua [Jesus] as did some of the other disciples. Now the Companion was known to the high priest and entered in with Yeshua into the court of the high priest..."37:2 "They led Yeshua therefore from Caiaphas into the Praetorium. It was early, and they themselves did not enter into that place, that they might not be defiled but might eat of the Pesach feast. With them also, standing nearby, were Miryam the mother of Yeshua, Ya'akov [meaning James – Mary Magdalene's younger brother], Yosef and Salome; Miryam, called the Migdalah, the Beloved Companion; and the disciple called Levi. They heard and saw all that occurred and, as others have testified to the same, then the law says that their testimony is true." 38:1

Yeshua answered, "My Kingdom is not of this world. If my Kingdom were of this world, then my people would fight, that I would not be delivered to the enemy. But now my Kingdom is not from here." Pilate therefore asked him, "Are you a King then? And Yeshua answered, "You say that I am King. For this reason I have been born, and for this reason I have come into the world that I should testify to the truth. Everyone who is of the truth listens to my voice." 38:4

[This next section is following the crucifixion of Jesus, as told in the Gospel of Mary Magdalene]

"All the disciples had gathered at the end of the week, when all the disciples had gathered in the house at Bethany, the Midgalah [Mary Magdalene] came to them and told them what she had seen and what Yeshua had said. But they were grieved and wept greatly, saying, "How shall we go out and preach the gospel of the kingdom of the Son of Humanity [Jesus]? If they did not spare him, how will they spare us?" 41:2

"Then the Midgalah stood up, greeted them all and, raising her right hand, said to her brethren, "Only from the truth I tell you, do not weep and do not grieve or be irresolute, for his grace and that of the one who sent him will be entirely with you and will protect you. But rather, let us praise his greatness, for he has prepared us and made us truly human." 41:3

"When the Migdalah said this, she turned their hearts to the good, and they began to discuss the words of Yeshua."41:4 Shimon Kefa said to the Migdalah, "Sister we know that he [Jesus] loved you more than any other among women, tell us the words of the teacher, which you remember, which you know and understand, but we do not, nor have we heard them." 41:5

"The Migdalah answered and said, what is hidden from you, I will proclaim to you. And she began to speak to them the words that Yeshua had given her." 42:1

[In the below section lies the truth of her ascension to Christ Consciousness, so lovingly taught to her by Jesus.]

"My master [Jesus] spoke thus to me. He said, Miryam, blessed are you who came into being before coming into being [a reference to the soul and reincarnation], and whose eyes are set upon the Kingdom, who from the beginning has understood and followed my teachings. Only from truth I tell you, there is a great tree [metaphor for the spine] within you that does not change, summer or winter, and its leaves do not fall. Whosoever listens to my words and ascends to its crown [top of the head] will not taste of death, but know the truth of eternal life." 42:2

"Then He showed me a vision in which I saw a great tree that seemed to reach unto the heavens [the flow of divine energy, the life-force, from the body to the heavens]; and as I saw these things, he said, 'The roots of this tree are in the Earth, which is your body, the trunk extends upward through the five regions of humanity to the crown, which is the kingdom of the Spirit." 42:3

"There are eight great boughs upon this tree and each bough bears its own fruit, which you must eat in all its fullness [meaning in each energy center you must release the negative emotion which keeps the energy dormant and like a domino affect it rises through the body up the spine to be release your life-force to God]. As the fruit of the tree in the garden caused Adam and Chav'vah [meaning life-giver or the first woman] to fall into darkness, so this fruit will grant to you the light of the Spirit that is eternal life. Between each bough is a gate and a guardian [metaphor for negative energy as witnessed through the illusion perpetuated by Lucifer-Satan] who challenges the unworthy who try to pass." 42:4

"The leaves at the bottom of the tree are thick and plentiful, so no light penetrates to illuminate the way. But fear not, for I am the way and the light, and I tell you that, as one ascends the tree [spine], the leaves that block one from the light are fewer, so it is possible to see all more clearly. Those who seek to ascend must free themselves of the world [material illusion]. If you do not free yourself from the world, you will die in darkness that is the root of the tree [survival instinct]. But if you free yourself, you will rise and reach the Light that is the eternal life of the Spirit." 42:5

"And as he said these things, I felt my Soul ascend and saw the First great bough that bears the fruit of love and compassion, the foundation of all things. And I knew that before you can eat of this fruit and gain its nourishment, you must be free of all judgment and wrath. When you have freed yourself of these burdens, you may eat of the fruit and so gain the love and compassion that will allow you to pass the first of the seven guardians. And I heard the voice of the Lord of wrath [Lucifer] calling to me, but I denied him and he had no part in me. 42:6

"I saw my Soul ascend again and he showed me the Second great bough, weighed down with the fruit of wisdom and understanding. And I saw that before you can taste of its bounty, you must be free of all ignorance and intolerance. Only then can you eat of the fruit of intolerance. Only then can you eat of the fruit and so pass upward unhindered through the second of Seven gates. And I heard the voice of ignorance call to me, but I knew him not, and so my Soul did thus unchallenged." 42:7

"Then my master showed me the Third great bough, which bears the fruit of honor and humility. Only when free of all duplicity and arrogance may you partake of its nourishment. And arrogance called me saying, "you are not worthy, go back." But my Soul was deaf to him, and so moved onward and upward into the increasing light. 42:8

"And then there came the Fourth bough, blossoming with the fruit of strength and courage. And I heard him tell me that to eat of this fruit, you must have freed yourself from the weakness of the flesh and confronted and conquered the illusion of your fears. And the Master of the world stood before me and claimed me as his own [Lucifer – the master of the material illusion and the creator God of the physical dimension who abandoned the true benevolent God above all dimensions], but I denied him and he had no part of me." 42:9

"Only then, my Master [Jesus] told me, when you have rejected the Deceiver [Lucifer-Satan], can you pass through the hardest gate [energy center] of all, to attain the Fifth bough and the fruit of clarity and truth. Only then will you know the clarity and truth of your Soul, and knowing yourself for the first time understand that you are a child of the living Spirit [God]. And as my Soul moved upward, I realized that I could no longer hear the voice of the world, as all had become as silence." 42:10

"Then in the light above, I saw the Sixth bough, the one that bore the fruit of power and healing. My Master [Jesus] told me that when you truly have eaten of the fruit of the clarity and truth of yourself, then could you partake of the fruit of power and healing, the power to heal your own Soul and thereby make it ready to ascend to the Seventh bough, where it will be filled by the fruits of light and goodness. 42:11

"And I saw my Soul, now free of all darkness, ascend again to be filled with the light and goodness that is the Spirit. And I was filled with a fierce joy as my Soul turned to fire [light] and flew upwards in the flames from whence my Master [Jesus] showed me the Eighth and final bough, upon which burned the fruit of the grace and beauty of the Spirit. 42:12

"And I felt my Soul and all that I could see dissolve and vanish in a brilliant light, in a likeness unto the sun, and in the light, I beheld a woman of extraordinary beauty [Mary saw her 'awakened' soul], clothed in garments of brilliant white. The figure extended its arms, and I felt my Soul draw into its embrace, and in that moment I was freed from the world, and I realized that the fetter of forgetfulness [meaning the unconscious, dormant state of the soul] is temporary. From now on, I shall rest through the course of time of the age in silence, and then as if from a great distance I heard the voice of my Master [Jesus] tell me, "Miryam, whom I have called the Migdalah, now you have seen the ALL [The "I AM" of Christ Consciousness], and have known the truth of your self; the truth that is 'I AM'. Now you have become the completion of completions." And thus, the vision ended. 42:13

"This is what my Master has told and shown me. And only from the truth I tell you, that all that I have revealed to you is true." 42:14

"When the Migdalah had told of all that Yeshua had said and done, she fell silent, since it was in that silence that Yeshua had spoken with her and revealed these truths." 42:15

"Many of the disciples did not understand what she had said, and grumbled against her among themselves." 43:1

"Andreas therefore answered and said to the brethren, "Say what you wish to say about what she has said. I at least do not believe that the Teacher said this. For these teachings are certainly strange and complicated ideas." 43:2

"Shimon Kefa answered and spoke concerning the same things, he questions them about Yeshua and said, "Did he really speak privately with this woman and not openly to us? Are we to turn about, and all listen to her? Did he prefer her to us?" 43:3

"Then the Midgalah wept and said to Shimon Kefa, "My brother Shimon Kefa, what do you think? Do you think that I have thought this up myself in my heart, or that I am lying about Yeshua? Only from the truth again I tell you that what I have said is the truth." 43:4

"And Levi answered and said to Shimon Kefa, "Shimon Kefa, you have always been hot-tempered. Now I see you contending against this woman like the adversaries. But if the Teacher made her worthy, who are you indeed to reject her? Surely as his companion, Yeshua knew her better than all others. That is why he loved her more than us." 43:5

"Rather, let us be ashamed and do as she says. Let us put on perfect humanity and acquire it as she has done, and separate as he commanded us and preach the testimony of the Son of Humanity, not laying down any other rule or other law beyond that which he gave us." 43:6

"And when they heard this, they were divided, and argued amongst themselves." 43:7

"These are the words and deeds of Yeshua, the "SON OF HUMANITY". There are also many other things which Yeshua did, which if they would all be recorded, then even the world itself would not have room for the books that would be written. I have testified and recorded all that I have seen and heard in the light and truth of his love and the grace and power of his word. Only from the truth I tell you, those amongst you who understand and believe his word will not know death." 44:1

"I am Miryam, called the Midgalah, the Beloved Companion." 44:120

—————•◆•—————

---

[20] The Gospel of the Beloved Companion: The Complete Gospel of Mary Magdalene, 2010, Jehanne de Quillan

The Seven Seals of Revelations

*"I AM the Alpha and the Omega, the First
and the Last, the Beginning and the End."*
                                    Revelations 22:13

With fresh eyes and an empty cup let us look at the true meaning behind Apostle John's words in the Book of Revelations and the opening of the Seven Seals. The true meaning of the opening of the Seven Seals has eluded the world until now. As you read the below words, the connection to John's Kundalini awakening in the metaphorical opening of the Seven Seals is remarkable.

John's message is clearly a warning to humanity to seek the Christ Consciousness by opening the seven energy centers (chakras) in the body, or be victim to the material illusion and the negative emotions that permeate in this dimension.

Historically, the Gospel of John was often in contradiction to the Gospel of both Thomas and Mary Magdalene albeit by intent or mistranslation, never-the-less early Christian leaders seeking to enforce a more orthodox and dogmatic approach to Christianity chose the Gospel of John to canonize rather than the Gospel's of Thomas and Mary Magdalene.

Despite their differences, John's Book of Revelation illustrates an awakening of his life-force energy, that was also experienced by Mary Magdalene, the difference being Mary's awakening was directly supported by Jesus and John's was towards the end of his life while in a cave on the island of Patmos.

The opening of the Seven Seals has always had the orthodox doomsday» interpretation, driven by those who have read John's words from the material illusion of the finite in an un-awakened state of being. However if you read John›s words as an illustration of Jesus' true message embodied in the Christ Consciousness and its attainment, you will see the metaphorical wonder of this story that has transcended history to captivate our minds.

As you will recall in our earlier discussion, a true definition of "apocalypse" in ancient Greek is to "uncover", to expose or a revelation.

Book of Revelation

John begins in chapter 4 with, "After this I looked, and behold, a door was opened in heaven..." The trumpets spoken of in revelations are the harmonic resonance (sound) when coupled with energy (light – the metaphorical fire of the Kundalini) sets the Soul free from the material illusion.

The First Seal – Characterized by deception

Vs 1 – "Now I saw when the Lamb opened one of the seals; and I heard one of the four living creatures saying with a voice like thunder, "Come and see."

Vs 2 – "And I looked, and behold, a white horse [truth]. He who sat on it had a bow; and a crown was given to him, and he went out conquering and to conquer."

The first energy center is attained through the harmonic resonance (sound) of the "Lam" or "Lamb". As you open this energy center, which is the rut of survival instinct at the base of the spinal cord, you are invited to "come and see" the slow initiation and release of the human vessel animal energies as seen in the lamb and horse, with both animals also being totems for innocence and truth.

The Second Seal – Characterized by war

Vs 3 – "And when he had opened the second seal, I heard the second beast say, "Come and see".

Vs 4 – "And there went out another horse that was red: and power was given to him that sat thereon to take peace from the earth, and that they should kill one another: and there was given unto him a great sword."

The second energy center is attained through the harmonic resonance of the "Vam" signifying a release of the life-force energy of power – the "great sword" representative of the masculine phallus. It is the center of control of the ego that drives humanity to kill one another and to conquer. The animal associated with this energy center is the crocodile.

The Third Seal – Characterized by famine

Vs 5 – "And when he had opened the third seal, I heard the third beast say, "Come and see". And I beheld, and lo a black horse; and he that sat on him had a pair of balances in his hand.

Vs 6 – "And I heard a voice in the midst of the four beasts say, A measure of wheat for a penny, and three measures of barley for a penny; and *see* thou hurt not the oil and the wine.

The third energy center is the harmonic resonance of the "Ram" and it corresponds to the digestive system, the conversion of food and liquids into energy and matter. The energies to defeat in this chakra are fear, anxiety, opinion and transition. It is the metaphorical 'measure of food and money' and defeating the innate fear represented by the "black horse".

The Fourth Seal – Characterized by pestilence

Vs 7 "And when he had opened the fourth seal, I heard the voice of the fourth beast say, "Come and see."

Vs 8 "And I looked, and behold a pale horse: and his name that sat on him was Death, and Hell followed with him. And power was given unto them over the fourth part of the earth, to kill with sword, and with hunger, and with death, and with the beasts of earth."

The fourth energy center is the harmonic resonance of the "Yam" and it corresponds to the heart and circulatory system. To attain the release of this powerful life-force energy, embodied in the heart, is to defeat the 'guardians of the gate' of the previous energy centers represented by the sword (phallus), hunger (digestive system) and death (survival animal instinct). The fourth energy chakra is characterized by projecting rays of light, 'behold the pale horse', to manifest healing and defeat the negative emotions in density.

The Fifth Seal – Characterized by tribulation

Vs 9 "And when he had opened the fifth seal, I saw under the altar the souls of them that were slain for the word of God, and for the testimony which they held:"

Vs 10 "And they cried with a loud voice, saying, How long, O Lord, holy and true, dost thou not judge and avenge our blood on them that dwell on the earth?"

Vs 11 "And white robes were given unto every one of them; and it was said unto them, that they should rest yet for a little season, until their fellow servants also and their brethren, that should be killed as they were, should be fulfilled."

The fifth energy center is the harmonic resonance of the "Ham" and it corresponds to the throat by which we speak our truth – the "word of God". Once the four previous energy centers are opened, and the negative dense energies (guardians of the gate) are overcome, you rise in your truth and the word of God transcending judgment and vengeance. As this chakra is opened and awakened, the physical illusion begins to fade away and there is an awareness of eternal presence, the metaphorical dawning of the "white robe".

The Sixth Seal – Characterized by a sign in the heavens

Vs 12 "And I beheld when he had opened the sixth seal, and lo, there was a great earthquake; and the sun became black as sackcloth of hair, and the moon became as blood;"

Vs 13 "And the stars of heaven fell unto the earth, even as a fig tree casteth her untimely figs, when she is shaken of a mighty wind."

Vs 14 "And the heaven departed as a scroll when it is rolled together; and every mountain and island were moved out of there places."

Vs 15 "And the Kings of the earth, and the great men, and the rich men, and the chief captains, and the mighty men, and every bondman, and every free man, hid themselves in the dens and in the rocks in the mountains:"

Vs 16 "And said to the mountains and rocks, Fall on us, and hide us from the face of him that sitteth on the throne, and from the wrath of the Lamb:"

Vs 17 "For the great day of his wrath is come; and who shall be able to stand?"

The sixth energy center is the harmonic resonance of the "OM", the most earth-shattering force of them all. It is the transcendental experience when you perceive the world not from the physical, but from the depths of your soul – the eternal presence. It can only be described as moving mountains, and shaking the very foundations of the physical realm. This divine energy center corresponds to the Third Eye located between the eyebrows. It is the mental projection of psychic powers, divine clarity and new realities. It is a mental projection into the sea of cosmic consciousness – the eternal "I AM".

The Seventh Seal – The connection to the divine Kingdom of God

Vs 1 "And when he opened the seventh seal, there was silence in heaven about the space of half an hour."

Vs 2 "And I saw the seven angels which stood before God; and to them were given seven trumpets."

Vs 3 "And another angel came and stood at the altar, having a golden censor; and there was given unto him much incense, that he should offer it with prayers of all the saints upon the golden altar which was before the throne."

Vs 4 "And the smoke of the incense which came with the prayers of the saints, ascended up before God out of the angel's hand."

Vs 5 "And the angel took the censor and filled it with fire of the altar, and cast it into the earth: and there were voices, and thunderings, and lightnings, and an earthquake."

Vs 6 "And the seven angels who had the seven trumpets prepared themselves to sound."

The seventh energy center is the attainment of the awakening and the divine connection to the kingdom of Heaven. In this moment, there is no harmonic resonance, merely silence – "there was a silence in heaven". It is as Mary Magdalene conveyed in her awakening. It is the final merging of your infinite energy field – your awakened soul – with universal consciousness, the Christ Consciousness. It is the condensing of the two into one, the male and the female aspects of your being merging to one, and thus residing in the eternal presence of God and the Angels.

And Yeshua said to his brethren, "Only from the truth I tell you, unless you change and become as little children, you will not gain the Kingdom. When you make the Two into One, and when you make the Inner like the Outer and the Outer like the Inner, and Upper like the Lower, and when you make Male and Female into a single one, so that the male will not be male nor the female be female, when you make eyes in the place of an eye, a hand in place of a hand, a foot in place of a foot, an image in place of an image, only then shall you gain the kingdom." *The Gospel of Mary Magdalene 30:12*

And the Brethren asked, "How does one who sees the vision see it—through the soul, or through the spirit?" The Savior [Jesus] answered and said, "One does not see through the soul, nor through the spirit, but the mind which is between the two: that is what sees the vision."

The Gospel of Thomas

————•◆•————

Closing Thoughts

"I only ask to be free. The butterflies are free."
Charles Dickens

I would like to leave you with a message of love and hope. Although we feel that we are just now discovering these innate, divine truths, if you quiet your mind and listen to your heart you will know that you are re-discovering yourself and your forgotten past.

This is the journey that we have made before, a journey we incarnated on Earth in this moment in time, to complete again. There is nothing to fear for you and all that you love is eternal, and lovingly protected by God.

We are the Sons and Daughters of Light who are re-claiming our truth and joining one another in the sea of collective consciousness. We are each a pearl strung along the necklace of infinity. We are the dream manifest by our creator – the God who resides above everything. We are the spectacular butterfly finally emerging from its cocoon. Celebrate life, celebrate love and celebrate your eternal existence. May God bless your journey home. **S**

A message of love and hope from Thoth, the Atlantean, in excerpts from the Emerald Tablet

"Man's evolution consists in the process of changing to forms that are not of this world. Grows he in time to the formless — to live on a higher plane. Listen, O man, to My voice, telling of pathways to Light, showing the way of attainment: how you shall become one with the Light: Seek ever more Wisdom.

Find it in the Heart of the Flame [light]. Know that only by your striving can Light pour into you. Only the one, who of Light has the fullest, can hope to pass by the guards of the Way, who prevent unworthy people from entering it. You shall cognize yourself as Light and make yourself ready to pass on the Way.

Wisdom is hidden in darkness. When shining with Soul-Flame, find you the Wisdom, then shall you be born again as Light, and then shall you become the "Divine Sun". Seek, O man, to find the great Pathway that leads to eternal Life — through the image of the "Divine Sun".... Know, O man, you are only a soul! The body is nothing! The soul is everything! Let not your body be a fetter.

Cast off the darkness and travel in Light. Learn to cast off your body, O man, and be free from it. Become the true Light and unite then with the Great Light. Know that throughout space the eternal and infinite Consciousness exists. Though from superficial knowledge It is hidden, yet still forever exists. The key to these Higher worlds is within you; it can be found only within. Open the gateway within you, and surely you, too, shall live the true life. Only the soul, which is free from the material world, has life, which is a really true life. All else is only a bondage, a fetter from which to be free. Think not that man is born for the earthly. Though born on the Earth, he is a light-like spirit.

… Darkness surrounds the souls seeking to be born in Light. Darkness fetters souls… Only the one who is seeking may ever hope for Freedom. The Great Light that fills all the cosmos is willing to help you, O man. Make you of your body a torch of Light that shall shine among men…. Hear and understand: the Flame is the source of all things; all things are its manifestation.

Know We that of all, nothing else matters except the growth one can gain with the soul. Know We the flesh is fleeting. The things men count great are nothing to Us. The things We expect from you are not of your bodies but are only the perfected state of the Souls.

When you can learn that nothing but progress of the Soul can count in the end, then truly you are free from all bondage, free to work in accordance with your predestination. Know, O man, you are to aim at Perfection, for only thus can you attain to the Goal! Know that the future is never in fixation but follows man's free will. Man can only "read the future" through the causes that bring the effects in destinies.

… Know you not that in the Earth's Heart is the source of harmony of all things that exist and have been on its face? By the soul you are connected with the Earth's Heart, and by your flesh — with the matter of Earth. When you have learned to maintain harmony in yourself, then shall you draw on from the harmony of the Earth's Heart. Exist then shall you while Earth is existing, changing in form, only when Earth, too, shall change: tasting not of death, but one with this planet, living in your body till all pass away.

… Three are the qualities of God in His Light-Home: **Infinite Power, Infinite Wisdom, Infinite Love.**

Three are the powers given to spiritual Masters: to **transmute evil,** to **assist good,** to **use discrimination.**

Three are the things They manifest: **Power, Wisdom, and Love.**

Three are the Manifestations of Spirit creating all things: **Divine Love** possessing the perfect knowledge, **Divine Wisdom** knowing all possible means of helping living beings in their development, **Divine Power** which is possessed by the Primordial Consciousness Whose essence is Divine Love and Wisdom. Darkness and Light are both of one nature, different only in seeming, for each arose from one Source. Darkness is chaos. Light is Divine Harmony. Darkness transmuted is Light.

This, My children, is your purpose in being: **Transmutation of darkness into Light!"**

————•+•————

Appendix:

Chakra System

Chakra is Sanskrit for "wheel, disk or center". It is perceived as a life-force energy emenating from the soul that receives, expresses and assimilates energy in the bodies bioenergetic sphere. The energy centers within the chakra system are located in major nerve gangalia locations along the spinal cord to the head.

The energy located in each of the seven centers acts as a divine mirror connecting the body, with the soul, and the seven dimensions of God's kingdom of light. Each chakra correlates to our consciousness and is represented by the cosmic Lotus flower that has a certain number of pedals, which contain different harmonic resonance (sound). The life-force energy is the convergence of light (energy) with harmonic resonance (sound) to connect and give birth to the eternal, infinite soul and the finite, material body.

The symbols denoting the specific energy centers (chakras) are as follows:[21]

---

[21] http://www.hindupedia.com/en/Chakra

FIRST CHAKRA

Location: Base of the spinal area of coccyx
No. of Petals: Four
Color: Red
Element: Earth
Sound: The "Lam" or "Lamb"
Governing Issues: Survival, security, safety, ability to be grounded to the material dimension.
Physical function: excretion and digestion, small intestine and colon; kidneys, sex glands/drive, reproduction, hips, legs, lower back, rectum, uterus.

SECOND CHAKRA

Location: Just above the genital area
Part nourished: Adipose tissue

No. of Petals: Six
Color: Orange
Element: Water
Sound: VAM
Governing Issues: Sexuality, self esteem, personal power; need to control, extent hold onto emotions or let go.
Physical function: Influences ovaries, uterus, fallopian tubes, pelvis, lumbar spine, kidneys, bladder, and large intestine. Center for cleansing, purification, health

THIRD CHAKRA

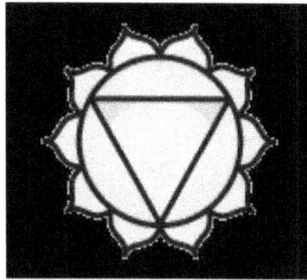

Location: Two fingers above the naval
Part nourished: Flesh and muscular system
No. of Petals: Ten
Color: Golden Yellow
Element: Fire
Sound: RAM
Governing Issues: When open, allows person to function normally even in time of distress, ability to connect, have long term relationships, love of family, home.
Physical function: Influences adrenal glands, profound effect on sympathetic nervous system, muscular energy, heartbeat, digestion, circulation, mood

FOURTH CHAKRA

Location: Eighth cervical vertebra of spine opposite region of heart.

Part nourished: Circulatory system

No. of Petals: Twelve

Color: Green

Element: Air

Sound: YAM

Governing Issues: Move beyond diaphragm, from outer courtyard to inner of body temple. Begin to recognize "self" is beyond definition, a source of light, love, human and divine. Allows one to sympathize with vibrations of other astral entities so can instinctively understand energies and atmospheres; can project rays of healing, manifest miracles; and is the gateway to astral body and hence regulator of emotional life.

Physical function: Influences thymus, located in center of chest behind upper breast bone, whose main function is proper utilization of amino competence factor which helps create immunity to disease.

FIFTH CHAKRA

Location: Base of neck by third cervical vertebra, just below throat by Adams apple.

Part nourished: Skin

No. of Petals: Sixteen

Color: Bright Blue

Element: Ether (the substance that connects energy)

Sound: HAM

Governing Issues: When open, you become aware of your mental body, and detach from the physical illusion. You have the awareness that internal worlds are real.

Physical function: Influences thyroid balance of entire nervous system, metabolism, muscular control and body heat production.

SIXTH CHAKRA

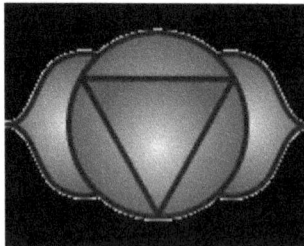

Location: Between eyebrows called The Third Eye

Part nourished: Bone marrow and production of RBCs

No. of Petals: Two
Color: Indigo
Element: Light
Sound: OM
Governing Issues: See clairvoyantly, communicate telepathically, heal through mental projection, can create new realities consciously to conform to unconditional joy.
Physical function: Pituitary master control center of mind/body affects all other endocrines.

SEVENTH CHAKRA – The Flower of Life

Location: At the Crown of head
Part nourished: Male and female reproductive principles
No. of Petals: Thousand
Color: Violet and White
Element: none
Sound: none
Physical function: Pineal gland
Governing Issues: The merging of the personal energy field with universal consciousness, and the merging of the feminine and masculine principles, eternal presence, to attain cosmic consciousness and enlightenment.

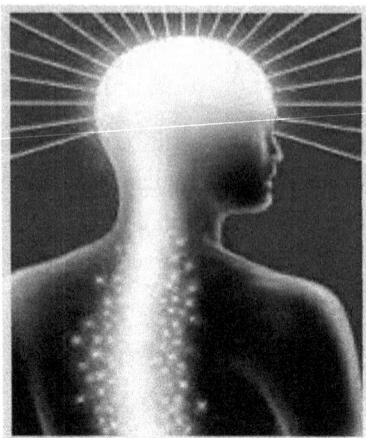

APPROACHING SINGULARITY

Bibliography

Agape Bible Study of the New Testament, http://www.agape-biblestudy.com/charts/The%20James%20of%20the%20New%20Testament.htm

Allen, J.H. *"Judah's Sceptre and Joseph's Birthright"; 1902*

Ammon-Wexler, Dr. Jill. "Pineal Gland & Your Third Eye: Develop "Conscious Self" Psychic Abilities"; 2011

Anonymous. "The Hermetic Arcanum: The Secret Work of the Hermetic Philosophy"; date unknown

Blavatsky, H.P. "Isis Unveiled"; 1877

Clow, Barbara Hand. "Alchemy of Nine Dimensions: The 2011/2012 Prophecies and Nine Dimensions of Consciousness"; 2004

Deeds of the Divine Augustus – emperor's autobiographical Res Gestae composed shortly before his death in 14 CE for posting on public monuments (from Internet Classics Archive at MIT). http://virtualreligion.net/iho/augustus.html

Ellis, Ralph. "Eden in Egypt"; Edfu Books, November 25, 2008; "Jesus, Last of the Pharaohs"; Edfu Books 1998

Faust, Michael. "Kabbalah, Hermeticism and the M-Theory"; Hyper-reality Books, 2010

Godwin, Joscelyn. "Atlantis and the Cycles of Time: Prophecies, Traditions and Occult Revelations"; 2011

Hall, Manly P. "The Secret Teachings of All Ages"; 1928

King James Bible, Oxford, (1769)

Kreisbert, Glenn. "Lost Knowledge of the Ancients"

Maat Sofia, The Ten Keys: Ancient Egyptian Roots of thePrincipia-Hermetica, http://maat.sofiatopia.org/ten_keys.htm

Macurdy, Grace Harriet, Phd. *"Vassal Queen and some contemporary women in the Roman Empire"*; Baltimore The Johns Hopkins Press; 1937

Martinez, Dr. Susan B. "Time of the Quickening: Prophecies for the Coming Utopian Age"; 2011

Mead, G.R.S. *"Thrice Greatest Hermes: Studies in Hellenistic Theosophy and Gnosis"*; York Beach, Maine: Samuel Weiser, 1992 [1906]

Melchizedek, Drunvalo. "Serpent of Light: The Movement of the Earth's Kundalini and the Rise of the Female Light 1949 to 2013"; Red Wheel/Weiser, LLC, 2007

Morrell, Peter. *"The Path of Non-Attachment"*; The Dalai Lama at Harvard, 1988

Naga, Jim. "How to Open Your 3rd Eye"; Mes Tree Publishing Press, 2010

Notable quotes & quotations http://quotes.liberty-tree.ca/quote_blog/Nathan.Mayer.Rothschild.Quote.4F94

Notovitch, translated, *"Life of Saint Issa, Best of the Sons of Men"*; 1894

Pagels, Elaine. "Beyond Belief: The Secret Gospel of Thomas"; Random House

Paracelsus, Philippus Aureolus Theophrastus Bombastus von Hohenheim. "The Book of the Revelations of Hermes: Concerning the Supreme Secret of the World"; 1493

Pope, H. *"Mary of Cleophas"*; 1910; In The Catholic Encyclopedia. New York: Robert Appleton Company. Retrieved February 20, 2012 from New Advent

Powell, Dr. Robert. "The Mystery, Biography & Destiny of Mary Magdalene"; Lindisfarne Books, 2008, "Prophecy, Phenonena, Hope"; Lindisfarne Books, 2011

Prophet, Elizabeth Clare. "The Lost Years of Jesus: Documentary

Evidence of Jesus' 17-year Journey to the East"; Summit Publications, Inc. 1984

Quillan, Jehanne de. "The Gospel of the Beloved Companion: The Complete Gospel of Mary Magdalene"; Athara Edition, 2010; NIV 1973

Ra, Summum Bonum Amen. "Summum: Sealed Except to the Open Mind"; 1975

Ramacharacka, Yogi. "The Collective Works"; Yogi Publication Society, 1908

Robinson, James M. ed. *"The Sophia of Jesus";* The Nag Hammadi Library, revised edition. HarperCollins, San Francisco, 1990

Roberts, J.M. *"Antiquity Unveiled";* 1892

Sams, Gregory. "Sun of God: Discover the Self-Organizing Consciousness that Underlies Everything"; 2009

Schiff, Stacy. *"Cleopatra: A Life";* Back Bay Books, 2010

Still, Bill. "No More National Debt"; 2011

Thoth, the Atlantean. "The Emerald Tablets of Thoth-the-Atlantean"; translations and intrepration by Doreal 1993

Three Initiates. "The Kybalion: A Study of the Hermetic Philosophy of Ancient Egypt and Greece"; The Yogi Publication Society, 1908

Tresmigistus, Hermes. "The Emerald Tablet of Hermes"; History of the Tablet, Holmyard 1957 and Needham 1980, "The Divine Pyramander" with translation by John Everard; original text 1500 AD

Vardaman, Dr. E. J., and Yamauchi, Edwin *M.; Chronos, Kairos, Christos: Nativity and Chronology Studies Presented to Jack Finegan,* Eisenbrauns 1989

www.ingramcontent.com/pod-product-compliance
Lightning Source LLC
Chambersburg PA
CBHW071621040426
42452CB00009B/1424